Ecological Studies

Analysis and Synthesis

Edited by

W. D. Billings, Durham (USA) F. Golley, Athens (USA)
O. L. Lange, Würzburg (FRG) J. S. Olson, Oak Ridge (USA)

Volume 30

Manfred Kluge · Irwin P. Ting

Crassulacean Acid Metabolism

Analysis of an Ecological Adaptation

With 112 Figures

Springer-Verlag Berlin Heidelberg New York 1978

Professor Dr. M. KLUGE
Institut für Botanik, Technische Hochschule
Schnittspahnstraße 3–5, 6100 Darmstadt/FRG

Professor I. P. TING
Department of Biology, University of California
Riverside, CA 92521/USA

ISBN 3-540-08979-9 Springer-Verlag Berlin Heidelberg New York
ISBN 0-387-08979-9 Springer-Verlag New York Heidelberg Berlin

Library of Congress Cataloging in Publication Data. Kluge, Manfred, 1936-. Crassulacean acid metabolism. (Ecological studies; v. 30). Bibliography: p. Includes index. 1. Crassulacean acid metabolism. 2. Acclimatization (Plants). 3. Botany—Ecology. I. Ting, Irwin P., joint author. II. Title. III. Series. QK881.K56.582'01'33.78-12658.

Typesetting, printing, and binding: Brühlsche Universitätsdruckerei, Lahn-Gießen.
2131/3130—543210

Acknowledgments

The preparation of this monograph would not have been possible without our many colleagues, associates, and students who contributed so much to our understanding of the physiology and ecology of Crassulacean Acid Metabolism plants. We are most grateful to Ph. Chimiklis, Z.Hanscom, O. L. Lange, U. Lüttge, R. Marcelle, S. K. Mukerji, T. Neales, C. B. Osmond, O. Queiroz, B. Sutton, and S. Szarek who contributed directly. Our colleagues P. N. Avadhani, W. Laetsch, W. W. Thomson, D. J. v. Willert, and K. Winter were generous in allowing us to use unpublished photographs and micrographs. We also thank Rosalia Heger, Doris Schäfer, and Fanita Terry for preparation of many figures throughout the text. Further, we must express our appreciation to the Deutsche Forschungsgemeinschaft and the U. S. National Science Foundation for financial support of our own personal research programs on CAM. The Botanic Gardens at the University of California, Berkeley, has been most generous in allowing the use of facilities and access to succulent plants. Our manuscript, of course, was not complete until typed by Inge Hill and Irene Schmidt, to whom we are grateful. Finally, we must thank Konrad Springer for his help and encouragement during the preparation of the manuscript.

December, 1978

MANFRED KLUGE
IRWIN P. TING

Contents

List of Abbreviations and Symbols

ABA = abscissic acid
Ala = alanine
AMP, ADP, ATP = adenosine mono-, di-, triphosphate
Asp = aspartate
CAM = Crassulacean Acid Metabolism
$\delta^{13}C$ = $^{12}C/^{13}C$ ratio, index of carbon isotope composition in plant material
DAAP = dihydroxyacetone phosphate
dm = decimeter
1,3 DPGA = 1,3 diphosphoglyceric acid
dwt = dry weight
F-6-P = fructose-6-phosphate
FDP = fructose-1,6-diphosphate
fwt = fresh weight
GAP = glyceraldehyde-3-phosphate
α-D-glu. = α-D-glucose
glu = glutamate
G-1-P = glucose-1-phosphate
G-6-P = glucose-6-phosphate
IRGA = infrared gas analyzer
J = joule
K_m = Michaelis constante
Mal = malate
MDH = malate dehydrogenase
n = index of cooperativity in the Hill equation
NAD^+, $NADH_2$ = nicotinamide adenine dinucleotide (oxidized, reduced)
$NADP^+$, $NADPH_2$ = nicotinamide adenine dinucleotide phosphate (oxidized, reduced)
NADP-TPDH = triosephosphate dehydrogenase (NADP dependent)
OAA = oxalacetate
α-OG = α-oxoglutarate
P_i = inorganic phosphate
PP_i = pyrophosphate
Ψ = water potential
PEP = phosphoenol pyruvate
PEP-C = phosphoenolpyruvate carboxylase
2-PGA = 2-phosphoglyceric acid
3-PGA = 3-phosphoglyceric acid
PFK = phosphofructokinase
Pyr = pyruvate
r = resistance to gas transfer
RudP = ribulose-1,5-diphosphate
RQ = respiratory quotient
S = index of succulence
S_m = index of mesophyll succulence
SWC = soil water capacity
TR = transpiration ratio
V_{max} = maximum velocity of enzymatically catalyzed reactions
VDP = vapor pressure deficit

Introduction

The acid metabolism of certain succulent plants, now known as Crassulacean Acid Metabolism (CAM) has fascinated plant physiologists and biochemists for the last one and a half centuries. However, since the basic discoveries of De Saussure in 1804 that stem joints of *Opuntia* were able to remove CO_2 from the atmosphere during the night, and of Heyne in 1815 (see Wolf, 1960) that organic acids accumulate in the leaves of *Bryophyllum calycinum* during the night, the two main aspects of CAM, diurnal CO_2 gas exchange and metabolism of malic acid, have first been studied nearly independently. Hence, it is not surprising that most research to elucidate the mechanism of CAM has been during the last 15 years since CO_2 exchange and malate metabolism were studied and interpreted in its context. These efforts finally resulted in a clear realization that the CAM phenomenon is a variation on the mode of how plants can photosynthetically harvest CO_2 from the atmosphere.

The interpretation of CAM in this sense was stimulated by the discovery of another variant of photosynthesis, the C_4-pathway (see Black, 1973; Hatch and Slack, 1970; Hatch, 1976). Because this newly discovered photosynthetic pathway is recognized to be very closely related to the CAM pathway, the work on the latter became intensified during these last years. The increased effort in research on CAM provided a large amount of new data which can now be collected and critically evaluated. We have undertaken this as one of the goals of the present monograph.

Furthermore, CAM is not merely an alternative biochemical variant of photosynthesis. Rather, the concept has emerged that CAM is a mechanism which facilitates a specialized group of plants to survive in habitats where water is temporarily deficient, and there are now compelling arguments to interpret CAM as an ecological adaptation. Unfortunately, among the considerable number of reviews on CAM resulting from the great tradition of research in this field, there is virtually none which in a satisfying manner pays attention to the ecological importance of CAM. To fill this gap, we feel encouraged to try the formulation of a more comprehensive ecological analysis of CAM in the background of known biochemical and metabolic aspects.

The outline of CAM provided here is not necessarily a complete documentation of historical developments in CAM research. These aspects have been extensively considered by Wolf (1960) and other authors (see below). Rather, our study will reflect our personal views and interpretations of the CAM phenomenon. We have written with the intention of stimulating further discussions and investigations on CAM which is a field of plant science where different disciplines such as taxonomy, morphology, anatomy, biochemistry, physiology, ecology, and plant geography cooperate in an ideal manner to increase the knowledge of how plants are able to conquer ecological niches.

The early history of CAM research is well documented in a variety of generally available reviews and in numerous theses which are not so available, but recommended for those with a research interest. Some major reviews are:

Evans (1932), Beevers et al. (1966), Bennet-Clark (1933a, b, 1937, 1949), Burris (1953), Davies (1959), Wolf (1960), Ranson and Thomas (1960), Walker (1962), Ting (1971, 1976), Ting et al. (1972), Black (1973), Osmond (1975, 1976a, 1978). Some theses of importance are: Bradbeer (1954), Haidri (1955b), Walker (1956a), Brown (1956), Avadhani (1957), Lyndon (1957), Stiller (1959), Champigny (1960), Khan (1969), Rouhani (1972), Morel-Claret (1973), Sutton (1974a), and Szarek (1974).

Terminology

Different terms have been used to describe the CAM phenomenon. In the English literature, the term Crassulacean Acid Metabolism is established and generally accepted because most species within the family Crassulaceae exhibit the phenomenon in a classical manner, and because Crassulacean plants were used in many of the initial fundamental studies. Unfortunately, the expression implies that CAM is restricted to the Crassulaceae, which is far from being true as will be shown later (see Chap. 1). In the German literature, often the term Diurnaler Säurerhythmus (i.e., Diurnal Acid Rhythm) is used, referring to the oscillation of malic acid content during the day/night cycle (see Wolf, 1960). This term has the advantage of neutrality in view of the taxonomic distribution of CAM plants, however, it does not exclude oscillations of acids other than malic which are not involved in CAM.

There are some reasons for using further the term Crassulacean Acid Metabolism (CAM). Firstly, it seems to be nearly impossible to find a term which characterizes the taxonomic distribution of CAM as well as the biochemical pecularities without being too unspecific. Secondly, the term Crassulacean Acid Metabolism is now so generally used that replacement would only cause confusion. Finally, it should not be forgotten that the term has an acronym (CAM) which is comfortable to read and speak.

Therefore, in this review we shall use the expression Crassulacean Acid Metabolism (CAM) to name the phenomenon under consideration. We regard plants as having CAM if they exhibit the following criteria which will be described later in more detail:

1. The malic acid content of the photosynthetic tissue must fluctuate in a diurnal rhythm with accumulation of malic acid during the night period and disappearance during the day.

2. The content of the storage carbohydrates of the photosynthetic tissues must fluctuate with malic acid, however, with inverse phase.

3. Normally, plants having CAM show net CO_2 uptake from the atmosphere during the night. During the day, a depression or cessation of net CO_2 uptake occurs. True CAM metabolism is accompanied by an inverse rhythm of stomatal opening with respect to non-CAM plants, i.e., stomatal opening occurs at night rather than during the daylight period.

It should be noted that neither morphological, nor cytological, nor taxonomical parameters are included in our definition of the term Crassulacean Acid

Metabolism. Throughout the text of this monograph, the basis for the above definition should become apparent.

Because of similar metabolic pathways in CAM and C_4 photosynthesis, frequently in the literature CAM plants are apostrophed as "C_4 plants." However, in order to avoid confusion, we suggest distinguishing between the "C_4 pathway of CO_2 fixation," "C_4 photosynthesis" and "CAM." The C_4 pathway of CO_2 fixation consists of the β-carboxylation of phosphoenolpyruvate and the conversion of the oxalacetate resulting from this reaction to C_4 acids such as malate or aspartate (for details see below). Hence, the C_4 pathway is part of both C_4 photosynthesis and CAM. However, the C_4 pathway may occur also in the photosynthetizing cells of C_3 plants (Krapf and Jacobi, 1975; Böcher and Kluge, 1977, 1978) or in nonautotrophic tissues (see Osmond, 1976a; Ting, 1971; Smith and Raven, 1976). The term "C_4 photosynthesis" should be applied to plants in which C_4 acids from the C_4 pathway are decarboxylated without being stored, and where the CO_2 from this decarboxylation enters the reductive pentose phosphate cycle. "C_4 plants" are plants having "C_4 photosynthesis." C_4 photosynthesis has been discussed in reviews by Hatch and Slack, 1970; Black, 1973; Hatch, 1976; Hatch and Osmond, 1976. Finally, CAM may be distinguished from C_4 photosynthesis by the fact that malic acid as a product of the C_4 pathway is temporarily stored before decarboxylation and further conversion by photosynthesis.

1. Taxonomy and Geographical Distribution of CAM Plants

Crassulacean Acid Metabolism has been documented in at least 18 flowering plant families (Table 1.1). As of 1977, there were recorded in the literature 109 genera and over 300 species of flowering plants (cf. Black and Williams, 1976; Szarek and Ting, 1977) with CAM. The largest and most important CAM families are the Cactaceae, the Crassulaceae, the Euphorbiaceae, the Aizoaceae, and perhaps the Liliaceae, especially if Agavaceae are included with the lilies. The epiphytic forms of the Bromeliaceae and the Orchidaceae make these families important.

Taxonomically these families do not show an obvious close relationship (Fig. 1.1). They do have, however, two distinct features in common. Firstly, all the families are tropical or subtropical in origin and many of the CAM species have invaded arid regions. The CAM species occupy arid or locally dry niches. These include arid desert-type habitats and physiologically dry habitats such as encountered by epiphytes and plants existing in very rocky or shallow soils. Some, in fact, such as *Dudleya* may actually grow within rock wall faces. Secondly, most of those species which show CAM are either stem or leaf succulents, or fleshy or coreiacous leafed plants such as the bromelid and orchid epiphytes (see Chap. 2.1).

1.1 Cactaceae

The Cactaceae are a moderately large family of about 100–150 genera and 1500 or more species (Britton and Rose, 1963). All are native to the Americas and are most abundant in the tropics (Fig. 1.2). Representatives extend as far north as British Columbia and they are native to every state in the U.S. except Alaska, Hawaii, Maine, New Hampshire, and Vermont. The genus *Rhipsalis* (Fig. 1.3) is naturalized in West Afrika, Madagascar, and Ceylon (Leese and Leese, 1959). There is extensive naturalization of *Opuntia* in South Africa and in the states of New South Wales and Queensland in Australia. As well, the Canary Islands have exotic *Opuntia* introduced from the Americas.

The family is divided into three subfamilies: the Pereskieae with a single genus *Pereskia*. The latter is unlike cactus in that fleshy leaves are persistent throughout the life cycle (Fig. 1.4). *Pereskias* can, nevertheless, be readily identified as cacti because of the presence of areoles, special stem structures giving rise to hairs,

Table 1.1. Flowering plant families known to have species showing CAM

Agavaceae	Euphorbiaceae
Aizoaceae	Geraniaceae
Asclepiadaceae	Labiatae
Asteraceae	Liliaceae
Bromeliaceae	Oxalidaceae
Cactaceae	Orchidaceae
Crassulaceae	Piperaceae
Cucurbitaceae	Portulacaceae
Didieraceae	Vitaceae

spines, and in some, ephemeral leaves. We have studied *Pereskia* and shown the presence of CAM by a marked diurnal fluctuation of organic acids.

The second subfamily, the Opuntieae, is characterized by the genus *Opuntia*, the chollas and prickly pears. The Opuntieae have about seven genera including *Pterocactus*, *Nopalea*, *Pereskiopsis* as well as *Opuntia*. Members of this subfamily are frequently branched and may have small, ephemeral leaves during the initial stages of growth (Fig. 1.5).

Opuntia has been most extensively studied. We have studied *Opuntia ramosissima*, *O. echinocarpa* (Ting and Dugger, 1968) and *O. basilaris* (Szarek et al., 1973). Also, *Opuntia acanthocarpa* Engelm. (Patten and Dinger, 1969). *O. echinocarpa*

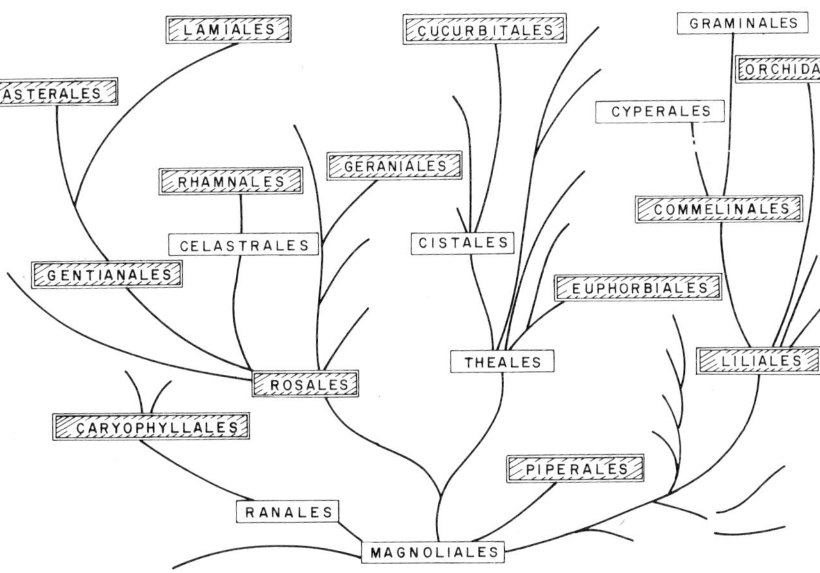

Fig. 1.1. Modified phylogenetic tree showing distribution of CAM families within orders. Tree drawn after Taktajan (1969). *Asterales* (Asteraceae); *Caryophyllales* (Aizoaceae, Cactaceae, Didiericeae, Portulacaceae); *Commelinales* (Bromeliaceae); *Cucurbitales* (Cucurbitaceae); *Euphorbiales* (Euphorbiaceae); *Gentianales* (Asclepiadaceae); *Geranales* (Geraniaceae, Oxiladaceae); *Lamiales* (Labiatae); *Liliales* (Agavaceae, Liliaceae); *Orchidales* (Orchidaceae); *Piperales* (Piperaceae); *Rhamnales* (Vitaceae); *Rosales* (Crassulaceae)

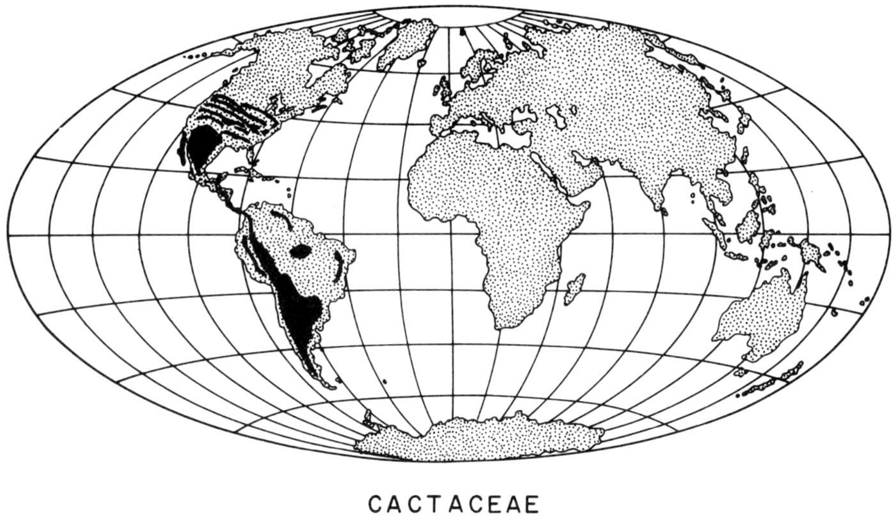

Fig. 1.2. Major geographical distribution of the *Cactaceae* (modified from Leese and Leese, 1959)

Fig. 1.3. *Rhipsalis sp.*, an advanced cactus distributed in tropical Americas and naturalized in Western Africa, Madagascar, and Ceylon

Fig. 1.4. *Pereskia sp.*, a primitive leafy cactus of the tropics. A member of the *Pereskieae*

Fig. 1.5. *Opuntia acanthocarpa* showing ephemeral leaves which appear for a short time during the spring

Fig. 1.6. The giant Saguaro (*Carnegia gigantea*) of the southwestern United States, one of the largest and most impressive CAM plants

(Ting and Dugger, 1966), *O. puberla* (Kausch, 1965), *O. inermis* (Osmond, 1975), and *O. versicolor* (Richards, 1915) have been investigated. Both dark CO_2 fixation and organic acid fluctuations confirm CAM metabolism in *Opuntia. Nopalea dejecta* was shown by Mukerji (1968) to be a CAM plant species.

The largest subfamily, the Cereae, has not been investigated exhaustively. This subfamily is characterized by the absence of leaves and a general columnar habitat so characteristic in the giant Saguaro (Fig. 1.6). The latter, *Carnegia gigantea*, was shown to have CAM by Despain et al. (1970) by demonstrating the typical inverted CO_2 fixation pattern. *Cereus* sp. (Milburn et al., 1968), *Echinocereus ledingii* (Patten and Dinger, 1969), *E. engelmanii* Rümpl (Ting and Dugger, 1968), *E. fendleri* Engelm., and *E. triglochidiatus* (Dinger and Patten, 1972), *Echinopsis eyriesii* Zucc. (Nuernbergk, 1961), *Ferocactus acanthodes* Britt. et Rose (Patten and Dinger, 1969; Nobel, 1977b), *Mammilaria tetrancistra* (Ting and Dugger, 1968), *M. rhodantha* Link et Otto (Nuernbergk, 1961), *Phyllocactus pfersdorffii*

Link (Nuernbergk, 1961), and *Zygocactus truncatus* Haw. (Nuernbergk, 1961; Holdsworth, 1971) have also been studied and shown to have CAM.

In all probability, all the Cactaceae have the CAM photosynthetic pathway.

1.2 Crassulaceae

Crassulacean Acid Metabolism has been reported and documented in at least nine genera and 28 species of the Crassulaceae. *Bryophyllum*, *Kalanchoe*, and *Sedum* are the genera most extensively studied. CAM has also been shown in *Aeonium haworthii* Webb et Berth. by Neales et al. (1968) using infrared CO_2 analysis, in *Cotyledon ladysmithiensis* (Schütte et al., 1967), and *C. peacockii* (Nishida, 1963) by acid fluctuations. In *Crassula* the species *Crassula macowaniana*, *Crassula multicana*, *Crassula argentea*, and *Crassula arborescens* have been studied (Schütte et al., 1967; Holdsworth, 1971; Milburn et al., 1968; Bonner and Bonner, 1948). CAM is known in *Dudleya lanceolata*, *D. saxosa* (Ting and Dugger, 1968), *D. farinosa* (Bartholomew, 1973), *Echeveria gibbiflora* (Holdsworth, 1971), and *E. kircheriana* (Nuernbergk, 1961). Nuernbergk (1961) demonstrated CAM in *Sempervivum tectorum* by dark CO_2 fixation.

In the genus *Kalanchoe*, *K. blossfeldiana* is studied most frequently. Interestingly, in var. Tom Thumb, CAM is only demonstratable when grown in short days (Queiroz, 1965). As well as *K. blossfeldiana*, *K. marmorata* (Milburn et al., 1968; Nishida, 1963), *K. welwitchii* (Nuernbergk, 1961), and *K. crenata* (Lyndon, 1962) are known to be CAM plants.

Of the *Sedum* species, *S. confusum*, *S. pachyphyllum*, *S. rubrotinctum*, and *S. spectabile* were shown by Holdsworth (1971) to have CAM, and *S. telephium* and *S. telephoides* were studied by Rouhani et al. (1973). *Sedum praealtum* (Fig. 1.7) was shown to have a distinct diurnal fluctuation of organic acids by Wood (1952). In *Sedum acre* CAM can be induced by stress (Kluge, 1977b).

Finally, the genus *Bryophyllum* has been extensively studied and shown to have CAM. *Bryophyllum daigremontiana* Berg. (= *Kalanchoe daigremontiana* Hamet et Perr.) (Fig. 1.8a) was studied by several authors, for example Holdsworth (1971), Nishida (1963), and Bonner and Bonner as early as 1948. *Bryophyllum tubiflorum* (= *Kalanchoe tubiflora*; see Fig. 1.8b) was among others investigated by Bruinsma (1958) and Kluge and co-workers. *Bryophyllum calycinum* was documented as a CAM plant by Milburn et al. (1968), Nishida (1963), and Pucher et al. (1949). *B. crenatum* was investigated by Milburn et al. (1968) and Bonner and Bonner (1948). The latter also studied *B. fedtschenkoi*.

The Crassulaceae are a reasonably large family with some 30 or more genera and over 1000 species. The largest genera are *Crassula*, *Sedum*, and *Kalanchoe*. Most are distinctly leaf succulents (Lawrence, 1955).

By and large, they are mostly northern hemisphere plants occurring principally in the arid regions of south central Asia, Mexico, the Mediterranean, but also extremely abundant in the tropics of South Africa (Fig. 1.9). There are virtually none in Oceania, Australia, and only a few species in South America.

In all probability, most of the Crassulaceae have CAM or at least may perform CAM under certain environmental conditions as shown for *Sedum acre* (Kluge, 1977b).

Fig. 1.7. *Sedum praealtum*, a typical crassulacean succulent

Fig. 1.8. a *Kalanchoe daigremontiana* (= *Bryophyllum daigremontiana* Berg), a typical broad-leafed CAM succulent of the Crassulaceae. **b** *Kalanchoe tubiflora* (= *Bryophyllum tubiflorum*). In this plant the photosynthetizing organs are represented by phyllodia

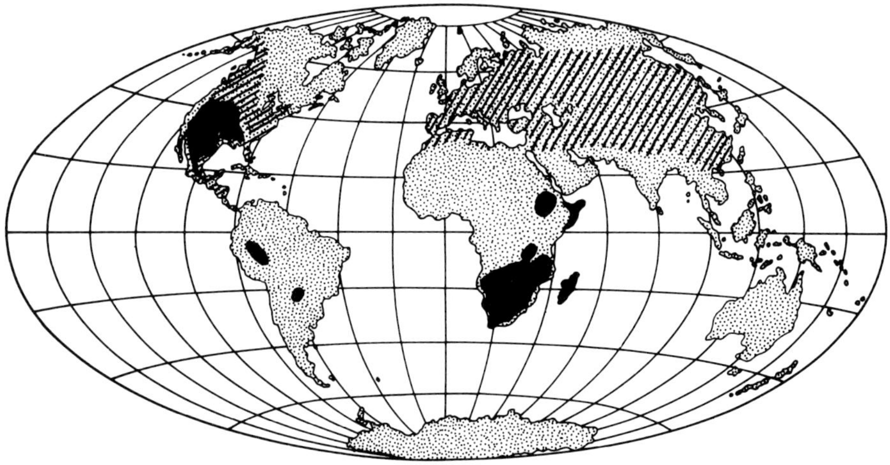

CRASSULACEAE

Fig. 1.9. Distributional map of *Crassulaceae* (modified from Leese and Leese, 1959)

1.3 Euphorbiaceae

The Euphorbiaceae are a large family of about 280 genera and some 7300 species (Lawrence, 1955). They can be either trees, shrubs, herbs, or vines and some are fleshily stemmed and cactus-like (Fig. 1.10). The major centers of distribution are tropical America and Africa. The succulent members are centered along both tropics in South America and Africa (Fig. 1.11) including the Canary Islands.

The succulent members of the genus *Euphorbia* have been most exhaustively studied. CAM has been shown in *Euphorbia submammillaris* and *E. xylophyloides* by dark CO_2 fixation and organic acid fluctuation (McWilliams, 1970). Nuernbergk (1961) showed dark CO_2 fixation in *E. grandidens*. Night-time organic acid accumulation was shown in *E. caducifolia* (Sen, 1970). McWilliams (1970) demonstrated dark CO_2 fixation and organic acid accumulation in *Monodenium lugardae*, *Synadenium grantii*, and *S. capulare*. It is reasonable to assume that all the succulent Euphorbiaceae have CAM.

1.4 Aizoaceae (Mesembryanthemaceae)

Crassulacean Acid Metabolism has been reported in at least seven genera of the Aizoaceae. Nuernbergk (1961) reported significant dark CO_2 fixation in *Gergeranthus multiceps* and in a *Faucaria* hybrid. Schütte et al. (1967) reported diurnal acid fluctuations in four South African aizoids: *Conophytum flavum*, *Lithops salicola* (Fig. 1.12), *Tetragonia fruticosa*, and *Trichodiadema barbatum*. *Lithops tubiniformis* and *L. insularis* were also reported to have Crassulacean Acid Metabolism (Holdsworth, 1971; Cockburn, 1974). Von Willert et al. (1977b) showed CAM in 24 of 27 Mesembryanthemacean species (see also von Willert et al., 1978).

Fig. 1.10. *Euphorbia grandidens*, a cactus-like succulent of Africa in the Euphorbiaceae which has been extensively studied as a CAM plant

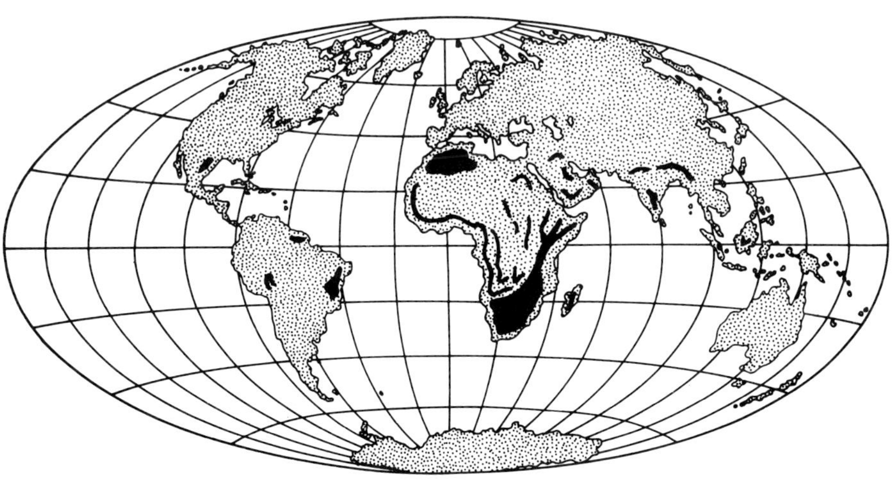

Fig. 1.11. Distributional map of the succulent CAM Euphorbiaceae (modified from Leese and Leese, 1959)

Fig. 1.12. *Lithops salicola*, an interesting African Aizoid known to show CAM

Fig. 1.13. *Conophytum megerae*, an African CAM Aizoid

Quite interesting is the report of Winter and v. Willert (1972; see also Winter and Lüttge, 1976) that *Mesembryanthemum crystallinum* shifts from a non-CAM metabolism to CAM when stressed with NaCl. In their experiments, only when irrigated with a 0.5 M NaCl solution or water stressed did the plant accumulate malate and fix measurably CO_2 in the dark (see also Chap. 3 and 6).

The family Aizoaceae (also referred to by various authors as Ficoidaceae or Mesembryaceae) is described in detail by Jacobsen (1960). It is a large family with over 100 genera and 600 species. By and large, the species are leaf succulents,

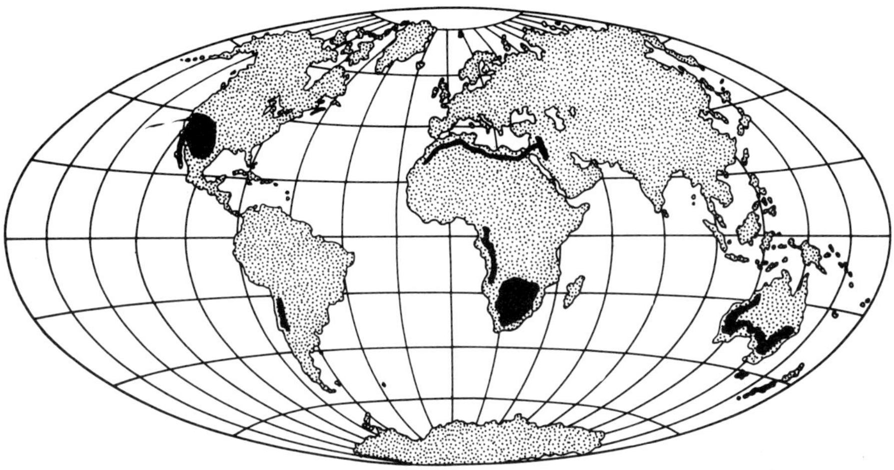

AIZOACEAE

Fig. 1.14. Distributional map of the Aizoaceae (modified from Leese and Leese, 1959)

although a few have much reduced leaves and are therefore better described as stem succulents (Fig. 1.13).

It is assumed that the center of dispersal of the Aizoid group is the generally frost-free region of southern Africa (Volk, 1960). The group extends into northern Africa, coastal Mediterranean, the deserts of the middle and near east, in California of the southwestern United States, New Zealand, Peru, and Australia. Generally, the distribution can be considered largely as Southern Hemisphere with the greatest development in the arid tropics and subtropics of South Africa and Western Australia (Fig. 1.14).

The anatomy of the leaves consists of relatively undifferentiated palisade-like mesophyll. In many cases there is a specialized, central water-storing tissue. Frequently the epidermis has large bladder-like cells resulting in the name "iceplant." Some species have large tannin sacs below the epidermis, and many contain much mucilage (Metcalfe and Chalk, 1950).

Metcalfe and Chalk (1950) consider the *Aizoids* to be closely allied to the Centrospermae.

1.5 Bromeliaceae

The Bromeliaceae are a large New World family of some 50 genera and nearly 2000 species. The family includes mainly thick, fleshy-leafed epiphytes inhabiting tropical and warm temperate America (Lawrence, 1955). A most notable exception to the epiphytic life form is *Ananas comosus*, the edible pineapple (Fig. 1.15). The most widely distributed genus is *Tillandsia.*

At least 14 genera and almost 50 species have been reported to have CAM. The most exhaustively studied genus is *Tillandsia* (Fig. 1.16) with at least 35 species shown to have CAM (Coutinho, 1969; Kluge et al., 1973; McWilliams, 1970; Medina, 1974; Medina et al., 1977).

Fig. 1.15. A pineapple field in the Hawaiian Islands. *Ananas comosus*, the edible pineapple, is the most important commercial CAM plant

Fig. 1.16. *Tillandsia usneoides* ("Spanish moss"), an epiphyte. This species represents a type of CAM plant which is not succulent in the usual sense

Other genera which have been shown to have CAM are *Acantostachis* (Coutinho, 1969), *Aechmea* (Coutinho, 1969; Medina, 1974), *Araeococcus* (McWilliams, 1970), *Billbergia* (Medina, 1974), and *Neoregelia* (Coutinho, 1963; McWilliams, 1970; Nuernbergk, 1961), *Bromelia* (Coutinho, 1964, 1969; Medina, 1974), *Canistrum* (Coutinho, 1963), *Dickya* (Coutinho, 1969; McWilliams, 1970; Medina, 1974), *Guzmania* (McWilliams, 1970; Medina, 1974), *Nidularium* (Nuernbergk, 1961), *Orthophytum* (Coutinho, 1969), and *Quesnelia* (Coutinho, 1963, 1969).

In the genus *Ananas*, as well as *Ananas comosus* (Neales et al., 1965; Medina, 1974), *A. sativus* (Milburn et al., 1968; Seshagiri and Suryanarayana-Murthy, 1951), *A. lucidus* (Medina, 1974), and *A. ananassoides* (Coutinho, 1969) have been shown to be Crassulacean Acid Metabolism plants.

Medina and Troughton (1974) and Medina et al. (1977) studied 87 species of 13 genera including representatives of the three subfamilies, Pitcairnioideae, Tillandsioideae, and Bromelioideae. They used techniques of dark CO_2 fixation, PEP-carboxylase assays, $\delta^{13}C$ values and organic acid fluctuation measurements to establish CAM. They found within the Pitcairnioideae, the most primitive group, that two genera out of six surveyed, *Dickya* and *Encholirion*, showed CAM. Of the Tillandsioideae, 19 species of 47 showed some CAM activity, and all the Bromelioideae studied (36 species) showed CAM. Of course, the inability to demonstrate CAM does not necessarily mean its absence in a group.

1.6 Asclepiadaceae

The Asclepiadaceae are a pantropical family most abundant in South America. A few members occur in temperate regions of both hemispheres. There are some 1800 species (Lawrence, 1955). Both leaf and stem succulents are known.

The fleshy genera *Hoya* (leaf succulent) and *Stapelia* (stem succulent) (Fig. 1.17) have been shown to have CAM. *Hoya carnosa* (Fig. 1.18) and *Stapelia variegata* have been shown to fix CO_2 in the dark (Nuernbergk, 1961) and *Stapelia nobilis* was shown to have CAM by diurnal organic acid fluctuations (Milburn et al., 1968). In *Frerea indica* the stems show CAM, however, not the leaves (Lange and Zuber, 1977).

The succulent *Asclepiadaceae* are largely distributed in Africa, East India, China, Malaya, and Australia.

1.7 Orchidaceae

The Orchidaceae are an extremely large family with some 450 genera and 10,000–15,000 species (Lawrence, 1955). They are herbaceous perennials, either terrestrial or epiphytic, and distributed throughout the globe. The epiphytic orchids, which are virtually all tropical, are the Crassulacean Acid Metabolism plants. These epiphytic orchids usually have coriaceous or succulent leaves.

The epiphytic orchids of tropical America have been most extensively studied by Coutinho (1963, 1965, 1969). He has demonstrated CAM by estimating diurnal pH changes in the following genera: *Brassavola* (1 species), *Cattleya* (11 species),

Fig. 1.17. *Stapelia gigantea*, a stem succulent of the Asclepiadaceae, native to Africa, is a CAM species

Fig. 1.18. *Hoya carnosa*, a leaf succulent of the Asclepiadaceae

Fig. 1.19. *Cattleya labiata*, an epiphytic CAM orchid

Encyclia (2 species), *Epidendrum* (3 species), *Laelia* (7 species), *Lanium* (1 species), *Oncydium* (1 species), *Schomburgkia* (1 species), *Sophronitis* (1 species), and *Vanilla* (1 species). Also, McWilliams has shown CAM by estimating dark CO_2 fixation and organic acid fluctuations in *Ascocentrum ampullaceum*, *Brassalaeliocattleya* sp., *Bulbophyllum gibbosum*, *Epidendrum alatum*, and *E. radicans*, *Phalaenopsis schilleriana*, and *Vanilla fragrans* (McWilliams, 1970).

Using an infrared gas analyzer, Nuernbergk (1961) showed dark CO_2 fixation in *Cattleya labiata* (Fig. 1.19), *Encyclia atropurpurea*, *Epidendrum schomburgkii*, and *Schomburgkia crispa*.

Knautt and Arditti (1969) using a cultivar of *Cattleya* measured $^{14}CO_2$ fixation into organic acids and Milburn et al. (1968) demonstrated CAM in *Epidendrum radicans* by diurnal fluctuations of organic acids.

1.8 Liliaceae

The Liliaceae are a large family of about 240 genera and 4000 species (Lawrence, 1955) widely distributed over the earth but especially abundant in warm, temperate, and tropical regions.

Many of the fleshy-leafed xerophytic members of this family have been shown to be CAM species. Four genera have been documented as CAM plants: *Aloe*, *Gasteria*, *Hawarthia*, and *Sansevieria* (Fig. 1.20). Dark CO_2 fixation was shown in *Aloe arborescens* (Fig. 1.21), *A. aristata*, *Gasteria verrucosa*, and *Sansevieria trifasciata* by Nuernbergk (1961). Changes in cell sap pH or organic acid fluctuations were shown in *Aloe globuligemma* (Holdsworth, 1971), *A. arborescens* (Denius and Ho-

Fig. 1.20. *Sansevieria zeylamica*, a CAM member of the Liliaceae

Fig. 1.21. *Aloe arborescens*, a very succulent-leafed lily

Fig. 1.22. *Yucca alata*, a CAM species of the Liliaceae, native to the southwestern United States

man, 1972), *Gasteria excelsa*, *Hawarthia zantneriana* (Schütte et al., 1967), and *Sansevieria liberica* (Milburn et al., 1968).

A diurnal fluctuation of organic acids and night stomatal opening can readily be measured in *Yucca* (Fig. 1.22) (Johnson and Ting, unpublished).

1.9 Agavaceae

The Agavaceae, frequently considered with the Liliaceae, are largely xerophytic, succulent-leafed perennials. *Agave americana* (Fig. 1.23) was clearly shown by Neales et al. (1968) to have an inverted CO_2 fixation pattern with uptake in the dark. Our own unpublished data with *Agave deserti* (Fig. 1.24) in southern California and those of Nobel (1976) show a marked diurnal fluctuation of organic acids, documenting CAM in Agavaceae.

The Agavaceae, including *Agave*, *Beschorneria*, *Calibanus*, *Dasylirion*, *Furcraea*, *Nolina*, and *Samuela*, are largely American-distributed from the southern United States into Mexico and northern South America and the West Indies.

Fig. 1.23. *Agave americana*, a CAM plant of the Agavaceae

Fig. 1.24. *Agave deserti*, a large CAM plant growing in its natural habitat of the southwestern United States

Fig. 1.25. An ornamental bed of *Kleinia*, a leaf-succulent CAM plant of the sunflower family

1.10 Asteraceae

The Asteraceae (Compositae) are a large family with nearly 1000 genera and 20,000 species ranging from annual herbs to trees and vines. The representatives of the family are widely distributed over the earth.

The succulent forms of the *Senecio* type are found largely in North and South Africa, the Canary Islands, East Indies, Madagascar, and Mexico.

Many succulent members (Fig. 1.25), e.g., *Aster*, *Kleinia*, *Natonia*, and *Senecio*, have been shown to have CAM.

Ganzmann and von Willert (1973) showed CAM in *Aster tripolium* by dark CO_2 fixation and organic acid accumulation. Schütte et al. (1967) measured diurnal organic acid fluctuations in *Kleinia tomentosa* and *Senecio cephalophorus*. Milburn et al. (1968) showed a diurnal fluctuation of organic acids in *Notonia petraea* and Thoday and Evans (1931) showed a fluctuation in *Kleinia articulata*.

1.11 Vitaceae

The Vitaceae are a tropical and subtropical family of 11 genera and about 600 species (Lawrence, 1955). They are mostly shrubs, but some are succulent in nature and two species of *Cissus* (Fig. 1.26), the largest genus, have been shown to have CAM. *Cissus hypoleuca* (Schütte et al., 1967) and *C. quadrangularis* (Milburn et al., 1968) were both shown to display large diurnal variations of organic acids. *Cissus* is native to Asia, tropical Africa, and arid Australia.

Fig. 1.26. *Cissus jutae*, a succulent member of the grape family (Vitaceae), known to show CAM

1.12 Geraniaceae

The Geraniaceae are a small family of eight genera and about 850 species (Lawrence, 1955) distributed in temperate and subtropical regions of both hemispheres. One species of *Geranium (Geranium pratense)* has been shown to have a diurnal fluctuation of organic acids (Thomas and Beevers, 1949) and hence, is suspected to have CAM. However, further investigations are desired. A diurnal fluctuation of organic acids has been observed in the succulent leafed *Pelargonium ceratophyllum* (Ting, unpublished).

Geranium pratense is native to Europe and northern Asia, and extends into tropical east Asia. The succulent *Pelargoniums* are largely African.

1.13 Other Families

In conjunction with the preparation of this monograph, several additional families suspected to have CAM by virtue of succulence were tested for diurnal fluctuation of acidity. Species growing under dry conditions in a glasshouse at the University of California Botanic Garden (Berkeley) were sampled in duplicate at dawn and again in late afternoon. Total acidity was titrated with 10 mM NaOH.

Fig. 1.27

Fig. 1.28

Fig. 1.27. *Xerosicyos*, a CAM succulent member of the Cucurbitaceae, native to Madagascar

Fig. 1.28. Several species of Didieriaceae, leaf and stem succulents of the Madagascar endemic family. The succulent leaves show CAM activity

The survey revealed the following new families as having probable CAM: Cucurbitaceae (Fig. 1.27), Didieraceae (Fig. 1.28), Labiatae, Oxalidaceae, Piperaceae, and Portulacaceae (Fig. 1.29). In addition, within the previously reported Geraniaceae, *Pelargonium ceratophyllum* was shown to have a diurnal fluctuation of acidity. These new data are summarized in Table 1.2.

It is our opinion that the species of Cucurbitaceae, Portulacaceae, Labiatae, and Didieraceae of arid Africa, Madagascar, and parts of the arid New World are valid CAM species (Fig. 1.30). The succulent leaf *Oxalis* species of South America and the *Peperomias* of Africa require more work.

Some nonflowering plants are reported to show CAM activity, perhaps the most interesting is *Welwitschia mirabilis* (Fig. 1.31), a member of the Gnetatae and hence one of the most advanced gymnosperms. Smith and Epstein (1971) reported a ^{13}C isotope composition suggestive of either C_4 or CAM activity ($\delta^{13}C = -14.4‰$; see Chap. 4). Dittrich and Huber (1974) and Schulze et al. (1976), subsequently reported other characteristics leading to the conclusion that *Welwitschia* is a CAM species but that members of the two other genera in the order, *Ephedra gerardiana* and *Gnetum gnemon* are not. These findings are most

Fig. 1.29. *Anacampseros*, a succulent genus of the Portulacaceae. The genus is native to Africa and Australia and known to show CAM

Table 1.2. Results of survey at the Botanic Garden, University of California, Berkeley

Family Species	Diurnal acid change [μeg g^{-1} fresh weight]
Portulacaceae	
Anacampseros albissima	99
Anacampseros sp.	136
Calandrina sp.	70
Talinum guadalupense	54
Didieraceae	
Alluaudia ascendens	225
A. humbertii	367
Didieria trolli	179
Cucurbitaceae	
Xerosicyos perrieri	93
X. danguyii	100
Labiatae	
Plectranthus prostratus	88
Oxalidaceae	
Oxalis carnosa	55
Geraniaceae	
Pelargonium ceratophyllum	77
Piperaceae	
Peperomia rankii	65

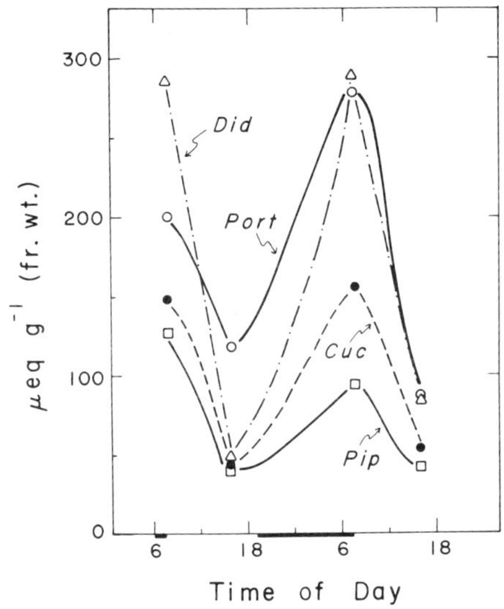

Fig. 1.30. Diurnal acid fluctuation in Didieriaceae (*Did* = *Alluaudia*), Portulacaceae (*Port* = *Anacampseros*), Cucurbitaceae (*Cuc* = *Xerosicyos*), and Piperaceae (*Pip* = *Peperomia*)

interesting since *Welwitschia* is the only gymnosperm known with characteristics of CAM. It is perhaps of significance that Jacobsen (1960) lists *Welwitschia* as a succulent plant.

Hew and Wong (Hew and Wong, 1974; Wong and Hew, 1976) have reported that two ferns, *Drymoglossum piloselboides* (L.) and *Pyrrosia longifolia* (Burm.), show dark CO_2 uptake, malic acid accumulation, and nocturnal stomatal open-

Fig. 1.31. *Welwitschia mirabilis*, a CAM-performing gymnosperm, at its natural stand in the Namib Desert (South West Africa) (Photo: von Willert)

ing. Thus, these epiphytic ferns with succulent fronds have the CAM characteristics.

In a preliminary communication, Gaur (1968) reported that *Salvadora persica* Lima (Salvadoraceae) and *Prosopis juliflora* D. C. showed a diurnal fluctuation of total titratable acidity. The presence of CAM in these species, however, is doubtful and interpretation of the data should await further experimentation.

1.14 Conclusions

The CAM plants found in the 18 families studied are largely distributed along the tropics of Cancer and Capricorn and in the wet tropics in physiologically dry, epiphytic type habitats. Habitat and niche rather than phylogeny appear to be important. Hence, convergent evolution is apparently the explanation for the existence of the CAM photosynthetic pathway in these plants.

The distribution within the families probably ranges from all members with CAM potential, as is the case of the tropical- and arid-distributed Cactaceae and Crassulaceae, to two genera out of about 20 in the Vitaceae and Geraniaceae. Of the genera occurring in each of the Vitaceae and Geraniaceae, i.e., *Cissus* and *Geranium (Pelargonium)*, both are xerophytic.

Of the large families with both CAM and non-CAM species, e.g., the Asclepiadaceae, Euphorbiaceae, and Liliaceae, the fleshy xerophytic forms have CAM. Of course, in the important CAM families, Bromeliaceae and Orchidaceae, most of the CAM species are epiphytes living physiologically dry niches.

Although the range of families, genera, and species investigated is not complete, it is somewhat unlikely that many new, important families will be found with extensive CAM representations.

An important problem which remains to be elucidated is the evolution of CAM. Is it truly an example of convergent evolution? The occurrence of CAM in the gymnosperm *Welwitschia* and in ferns suggests a positive answer to that question. If so, are the specific metabolic details of CAM in each family similar or are there distinctive differences which are representative of the taxonomic categories? In this context, did these plants with the CAM photosynthetic pathway actually all develop in the tropics and then move into arid environments of the subtropics and deserts of the world?

2. Morphology, Anatomy, and Ultrastructure of CAM Plants

2.1 What is a Succulent?

CAM is usually regarded as a typical feature of succulents because of its occurrence in many succulent species. However, two questions must be asked:

1. Is the occurrence of CAM restricted to succulents?
2. Do all succulents exhibit CAM?

To answer these two questions, the term succulence requires careful consideration.

The classification of a plant as succulent is based exclusively on morphological criteria, and does not implicate a special taxonomic status. The single morphological criterion which classifies a plant as succulent is the possession of voluminous water-storing tissues resulting in an increase in volume relative to surface area. Thus, a thick, fleshy, juicy habitus results which is envisaged as "succulence" [succus (lat.) = sap, juice], and which results in a form tending towards a spherical shape rather than disc shape typical for most leaves. Hence, succulents are generally characterized by their ability to store relatively large amounts of water.

There is no doubt that the storage of water by succulents represents an ecological adaptation, because succulents are xerophytes which occupy habitats where either precipitation occurs only periodically or soil water is generally not available (e.g., epiphytes) and periods of water deficiency are overcome with endogenous water reserves (see Chap. 6).

In succulents, all basic organs of the plant can function as water reservoirs. Thus, leaf succulents, stem succulents, and very rarely root succulents can be distinguished. CAM plants are only known as members of the leaf and stem succulents.

Storage of water in succulent plant organs occurs in the basic tissues. Those tissues, clearly having the task of storing water, are called water tissues. According to Haberlandt (1918), external and internal water tissues can be distinguished. External water tissues are modifications of the epidermis or subepidermis. They are characterized by very large cells which are often arranged in more than one layer. These cells either lack chloroplasts completely or at best have very few. External water tissues are exemplified by species of the genus *Peperomia* (Fig. 2.1; see Kaul, 1977). Internal water tissues can be derived from different types of parenchyma. In leaf succulents, the internal water reservoirs are provided by the mesophyll of the leaf. In stem succulents, the cortex, the pith, or both can contribute to the storage of water. In the case of water tissues derived from the inner part of the leaf mesophyll or the stem pith, they may be free of chloroplasts, as is the case for external water tissues. A typical example is given in Figure 2.2. Contrastingly, in succulents where the internal water tissues are identical with the photosynthetic parenchyma, the water-storing cells contain chloroplasts (Fig. 2.3). This is found either in those leaf succulents where the cells of the total mesophyll or at least of the outer parts of the mesophyll show both water storage and photosynthesis, and in those stem succulents where the photosynthesizing layers of the cortex participate in storing water.

2.2 Quantitative Indices of Succulence

Different indices have been proposed in the literature to describe quantitatively the capacity of a plant or a plant organ to store water, i.e., to quantify its succulence.

a) Surface Expansion

Succulence reflects a tendency for reducing the surface of a plant relative to its volume (see Figs. 1.6, 1.10). If we assume a perfect sphere (perhaps the ideal succulent form) with a total volume of 1000 cm^3, the diameter would be 12.4 cm with a total surface area of 484 cm^2. A disc-like leaf of 1000 cm^3 volume and 1 mm thick would have a total surface area over 20,000 cm^2. Hence, the disc shape will have a surface area some 41 times greater than the spherical shape of the same volume. Thermal consequences of the spherical shape for CAM plants will be discussed later (Chap. 5.3.6). The surface:volume ratio of a plant or a plant organ, as well as its water-storing capacity, is quantitatively described by an index called Oberflächenentwicklung (i.e., surface expansion; see Walter, 1926; Larcher, 1973)—where

$$\text{surface expansion} = \frac{\text{surface (cm}^2\text{)}}{\text{fresh weight (g)}}$$

Here, the fresh weight approximates the plant water content, assuming that 1 g fresh weight is equivalent to cm^3 since the plant volume is largely a function of the water content. Some computations of surface expansion are given in Table 2.1. As can be seen from the table, surface expansion is determined to some extent by the water status of the tissue (see Walter, 1926).

b) Degree of Succulence (S)

An alternative index of the water-storing capacity was introduced by Delf (1912). This index, termed degree of succulence (*S*), is

$$S = \frac{\text{saturating water content (g)}}{\text{surface (dm}^2\text{)}}$$

S values of typical succulents are in the range of 5.1–14.9 g dm^{-2} compared to nonsucculents with values of 0.1–1.2 g dm^{-2}.

Table 2.1. Surface expansion ($cm^2 \cdot g^{-1}$) in plants kept under dry and wet conditions (from Walter, 1926)

Species	Growth conditions		
	Normal	Rather wet	Very wet
Sedum dendroideum	10.0	18.2	22.7
Sedum altissium	13.5	30.8	—
Sempervivum assimile	11.1	17.5	27.0
Echeveria stolonifera	9.5	17.5	—
Mesembryanthemum curviflorum	12.8	16.7	—

2.3 Succulence and the Occurrence of CAM

2.3.1 Succulent CAM Plants

It should be mentioned that critical values either of surface expansion or of the degree of succulence (S) have not been defined which categorize a plant as being succulent.

Thus, the characterization of a plant as a succulent includes a certain degree of subjectivity. It is rather difficult to determine if all succulents show CAM, or to ascertain if some nonsucculents can perform the CAM carbon metabolism. Other difficulties arise from some recent discoveries (Winter and von Willert, 1972; see also Winter and Lüttge, 1976), that CAM is in some cases a rather flexible system. Its occurrence and the intensity of its occurrence seems at least in some species highly dependent on environmental conditions (see Chap. 5 and 6). Thus, failure to demonstrate CAM does not necessarily mean the absence of CAM in principle, but could mean its absence under the environmental conditions prevailing during the time of observation.

Based on what is now known about the CAM mechanism, three structural preconditions seem necessary (see also Nuernbergk, 1961). Firstly, the accumulation of substantial amounts of malic acid during the night, an essential feature of CAM, requires a compartment where the acid can be temporarily stored without endangering structures and biochemical processes of the cytoplasm. Such a compartment is provided by big vacuoles where malic acid is stored as an aqueous solution. Secondly, as will be shown later, photosynthesis is the essential pathway by which malic acid is consumed and transformed into carbohydrates during CAM. Thus, intact chloroplasts must be present. Thirdly, the intracellular sites of malic acid synthesis, malic acid storage, and malic acid consumption should presumably be localized in close vicinity. These three features occur only in those cases where a large vacuole and chloroplasts are localized in the same cells. Given these three structural preconditions, it is obvious that a high degree of succulence should correlate with CAM if for no other reason than a large storage capacity for malic acid. Numerous observations do indeed indicate a relation between succulence and the occurrence of CAM. For example, Nuernbergk (1961) and Queiroz (1966, 1968a, b, 1970) reported the occurrence of CAM only in the succulent leaves of short-day-treated individuals of *Kalanchoe blossfeldiana*. In contrast,

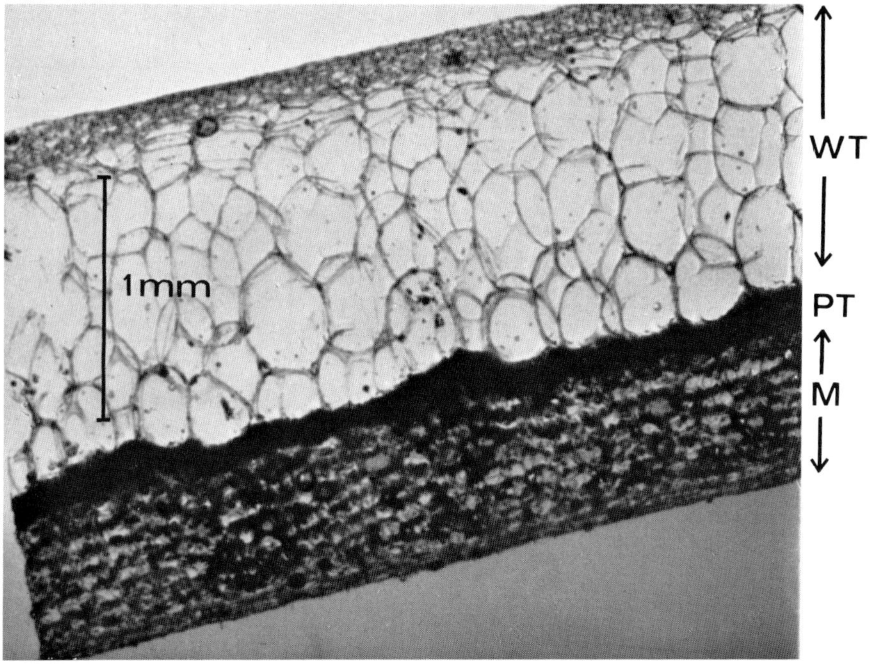

Fig. 2.1. Cross section through a leaf of *Peperomia obtusifolia* showing external water tissues. *WT*, water tissue; *PT*, palisade tissue containing chloroplasts without starch; *M*, starch- and chloroplast-containing mesophyll cells

CAM did not occur in the nonsucculent leaves of long-day-treated individuals of the same species. In *Hoya carnosa* (succulent) CAM could be observed, but CAM was not present in the nonsucculent *Hoya bella* (Nuernbergk, 1961). Another example is provided by the genus *Euphorbia*, where *Euphorbia grandidens* (succulent) is a CAM plant but not *Euphorbia pulcherrima* (nonsucculent; Nuernbergk, 1961). In *Kalanchoe tubiflora*, Winter (personal communication) observed CAM to occur only in the older (succulent) phyllodia and not in the young (nonsucculent) ones at the top of the plant. In *Frerea indica*, the succulent stem performs CAM; the nonsucculent leaves, however, do not (Lange and Zuber, 1977).

The structural preconditions outlined above imply as well that a high degree of succulence is not in itself a criterion for CAM. Only those succulents with big water-storing vacuoles accompanied by chloroplasts in the same cell fulfill these conditions (see Fig. 2.3a). This is the case in all the typical CAM plants now known. Consequently, we must exclude succulents as CAM plants, if water storage is *exclusively* in nongreen tissues, for example in exogenous water tissues or in chloroplast-free endogenous water tissues. This conclusion is confirmed by a number of observations. For example, neither *Peperomia*, a genus owing its succulence exclusively to a chloroplast-free exogenous water tissue (see Fig. 2.1) nor *Zygophyllum*, which has a nongreen endogenous water tissue sharply separated from the photosynthetic mesophyll, show all CAM activities in the sense of the above definitions (Fig. 2.2). *Ananas comosus*, which is clearly a CAM plant, has

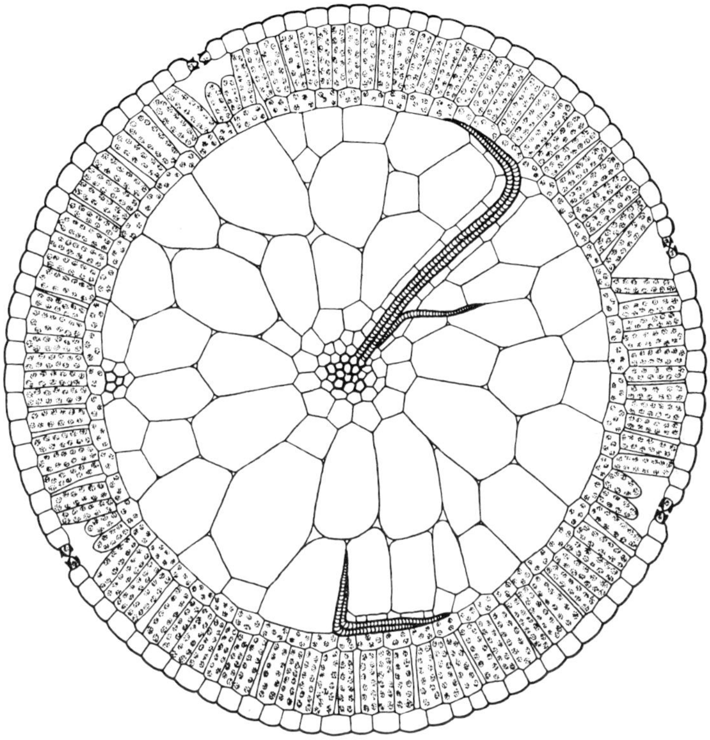

Fig. 2.2. Cross section through the leaf of *Zygophyllum simplex* showing a core of water tissue (internal water tissue) surrounded by palisade tissue. Tracheids connect the central bundle to the photosynthesizing palisade cells (from Stocker, 1956)

also an exogenous water tissue. However, unlike *Peperomia*, in *Ananas* the photosynthetizing cells also store water. Hence, *Ananas* is comparable to *Aloe arborescens* where the leaves, additionally to the CAM performing green tissue, have an endogenous water tissue (see below). Lüttge and Winter (personal communication) observed that in *Mesembryanthemum crystallinum*, the huge bladder cells localized on the surface of the epidermis do not accumulate malic acid even if the mesophyll of the same leaves showed CAM. Because it is reasonable to interpret the bladder cells of *Mesembryanthemum* as exogenous water tissues, these results show that the transport of malic acid from the site of synthesis to the site of storage is only effective within the distance of one or a few cells. Rapid transport from one tissue to the other, for example from the photosynthetic mesophyll to an exogenous or endogenous water tissue, seems to be too slow and ineffective for CAM (see also Chap. 4). Results of Sutton (unpublished), who showed that malic acid which was synthesized and stored in the cortical tissue of the CAM plant *Opuntia bigelovei* was not immediately transported into the central pith tissue, support the conclusion (Fig. 4.17). Similar results were obtained by Kluge and Knapp (unpublished) in *Aloe arborescens* (see also Denius and Homan, 1972) where CAM proceeds exclusively in the outer green tissue and not in the internal water tissue of the leaves. Malic acid synthesized during CAM remained in the outer green tissue rather than being transported to the water tissue. Thus, the following conclusion seems to be applicable:

Those succulents which unify the sites of malic acid synthesis, storage (large vacuoles), and conversion (chloroplasts) all within the same cells, can be expected to have CAM (see Fig. 2.3). In contrast, those succulents where the potential malic acid stores (water cells) are spatially separated from the sites of photosynthesis will not or only very weakly perform CAM, probably because of the complexity of transport of malic acid.

This interpretation of the relationship between the occurrence of CAM and plant succulence certainly requires more evidence. However, it agrees well with Haberlandt's (Haberlandt, 1918) concept, which interprets the evolution of anatomical structures in plants as a tendency to provide the shortest transport pathways. This principle is verified in the assimilating cells of CAM plants.

2.3.2 Nonsucculent CAM Plants

Further objections to the use of the term succulence when considering CAM are seen in the findings of Coutinho (1963, 1969) and Kluge et al. (1973), that *Tillandsia usneoides* (see also Medina et al., 1977) and other species of the genus *Tillandsia* are CAM plants. Normally, one would hesitate to classify *Tillandsia usneoides* as a succulent because of its low value of S (1.0–2.5 gm f.wt.dm^{-2}; compare typical succulents with 5.1–14.9 gm f.wt.dm^{2}) and its nonsucculent habit (Fig. 1.16). Hence, here it is tempting to assume also that nonsucculents may have CAM metabolism. However, micrographs of cross sections of *Tillandsia* leaves (Fig. 2.4) show mesophyll cells with relatively large vacuoles and few chloroplasts. Thus, these cells of *Tillandsia* strongly resemble the assimilatory cells of typical CAM plants.

The anatomical basis for the low succulence in *T. usneoides* is that there is only a relatively small number of mesophyll cells composing the leaf architecture. Hence, if the term succulence referred only to the capacity of photosynthetic cells to store water and malic acid, the single mesophyll cells in *T. usneoides* should be considered succulent. A relationship similar to *Tillandsia usneoides* can be expected in the other nonsucculent or weak-succulent species of the genus *Tillandsia* in which CAM could be demonstrated (Kluge et al., 1973; Medina et al., 1977). On the other hand, species having both a succulent habit and CAM can be found among the Tillandsioidea and other closely related Bromeliaceae (see Chap. 1). Thus, these taxa should provide good material to study the importance of succulence for CAM (see also McWilliams, 1970).

Madhusudana Rao et al. (1977) reported recently on the occurrence of CAM in non-succulent shrubs of the Indian desert. Further investigations of this phenomenon in particular with respect to interrelationship between CAM performance and cell anatomy are desired.

2.3.3 Mesophyll Succulence (Sm) as a New Index of CAM Capacity?

The occurrence of CAM in *Tillandsia usneoides* and other non-succulent *Tillandsia* species suggests that if the term succulence is to be useful in considering

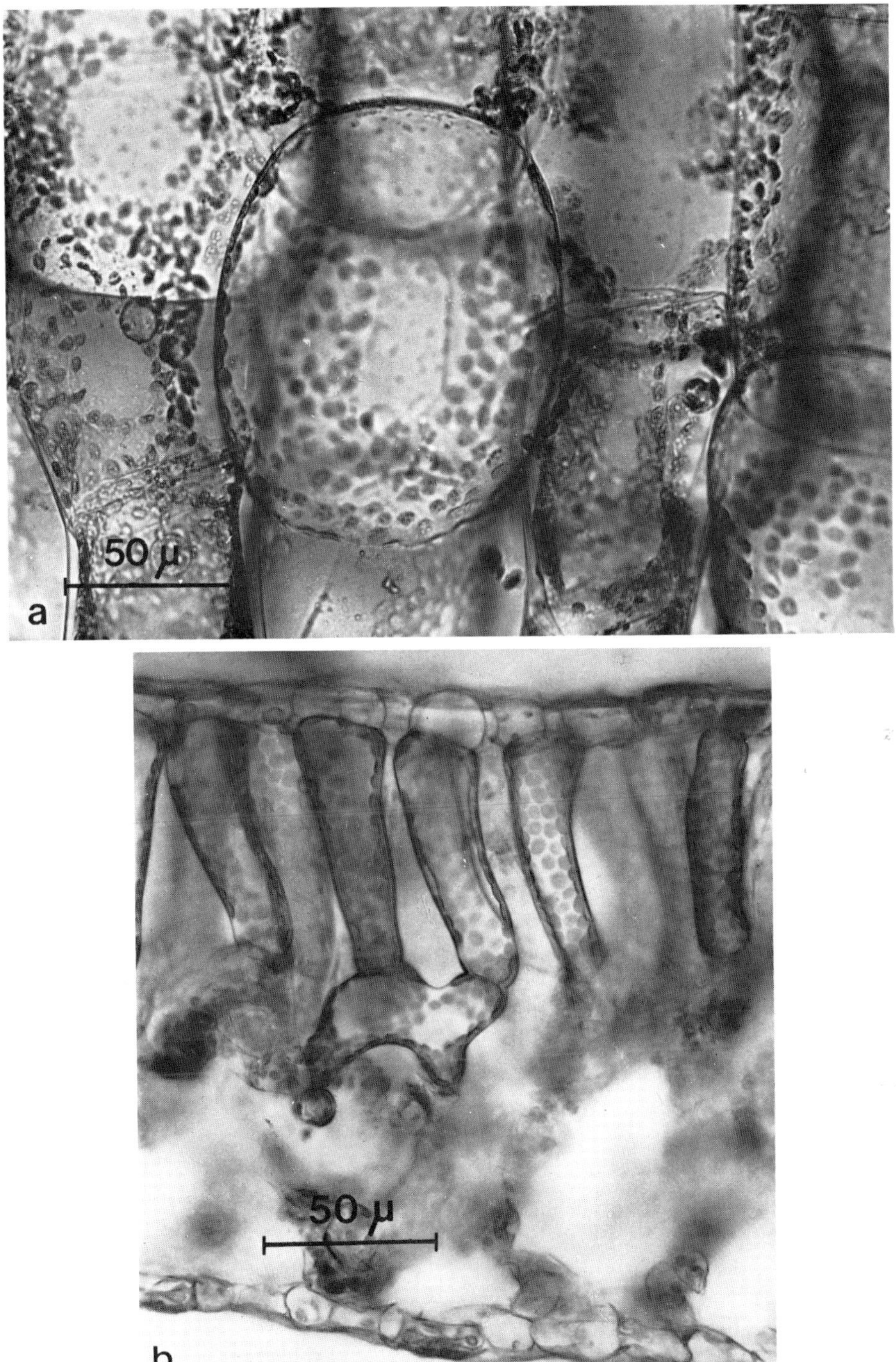

Fig. 2.3. a Mesophyll cells from a cross section through the leaf of *Kalanchoe daigremontiana*. Example of a water-storing, photosynthetic mesophyll typical of CAM tissues. **b** Mesophyll cells from a cross section through the leaf of *Vicia faba*. The photosynthetic cells of this C_3 plant are characterized by relatively small vacuoles and a high density of chloroplasts (from Kluge: Biologie in unserer Zeit. *4*, 120–129, 1972)

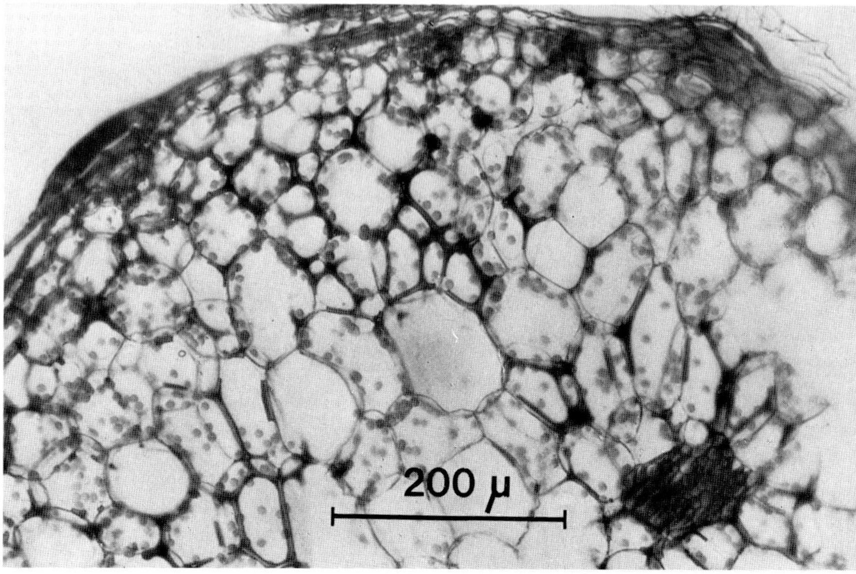

Fig. 2.4. *Tillandsia usneoides.* Cross section through a leaf showing cells of the CAM-performing mesophyll

CAM, it should be defined at the level of the photosynthesizing cells or tissues rather than at the level of the whole plant or plant organ.

In typical CAM cells (see Fig. 2.3a), a given number of chloroplasts (i.e., a given amount of chlorophyll) is in association with a very large vacuole (i.e., a large amount of water). In contrast, most photosynthesizing cells of non-CAM plants have relatively smaller vacuoles (Fig. 2.3b), and thus clearly less water per given amount of chlorophyll. Hence, it seems that the water content:chlorophyll content ratio of a tissue could provide a reasonable index of succulence at the cellular or tissue level. We suggest naming this index mesophyll succulence (*Sm*), where

$$Sm = \frac{\text{water content (gm)}}{\text{chlorophyll content (mg)}}$$

In those plant organs where nearly all cells have chloroplasts, this ratio can easily be measured. In other plants or plant organs having water tissues free of chloroplasts, the photosynthesizing tissues must be removed from the water tissues and analyzed separately. Confessedly, considerable technical difficulties may arise for this circumstance.

Some estimates of Sm values from a variety of plants are given in Table 2.2. It can be seen that if succulence is expressed on a chlorophyll base rather than on leaf area, *Tillandsia usneoides* fits well in the group of classical CAM plants such as *Kalanchoe daigremontianua* and *K. blossfeldiana.* Clearly, at the moment it is still too early to decide whether the introduction of Sm values provides an advantage in classifying plant species as candidates from CAM performance.

Table 2.2. Values of mesophyll succulence (*Sm*) in plants with and without CAM

Plant species	Mode of photosynthesis	*Sm*
Commelina africana	non-CAM	0.621
Zea mays	non-CAM	0.328
Impatiens sultani	non-CAM	0.776
Capsicum annuum	non-CAM	0.383
Geranium sp.	non-CAM	1.250
Cucurbita pepo	non-CAM	0.346
Coleus	non-CAM	0.620
Vicia faba	non-CAM	0.560
Gossipium turbareum	non-CAM	0.420
Kalanchoe daigremontiana	CAM	1.34
Kalanchoe tubiflora	CAM	8.40
Kalanchoe fedtschenkoi	CAM	2.15
Kalanchoe blossfeldiana	CAM	2.45
Sedum morganianum	CAM	13.00
Tillandsia usneoides	CAM	1.52
Carpobrotus edulis	CAM	3.90

2.4 The Presence of the Photosynthetic Apparatus as a Prior Condition for the Occurrence of CAM

CAM is a modification of carbon flow during photosynthesis (see Chaps. 3 and 4), thus integrating the photosynthetic machinery into its mechanism. Keeping this in mind, it can be predicted that CAM should occur only in those parts of a plant where the photosynthetic apparatus is fully developed (see Chap. 2.3.1).

No case has yet been reported where CAM has occurred in nongreen plants or plant organs. This supports the prediction that CAM depends strictly on the presence of chloroplasts. However, this generalization does not deny the possibility that portions of the metabolic sequence may occur in tissues or cells devoid of chloroplasts. Indeed, roots and other nongreen tissues have enzymes which are common to the CAM pathway, for example PEP carboxylase and malate dehydrogenase (Ting, 1971). Consequently, these nongreen tissues have the capacity to synthesize malate via β-carboxylation of PEP. Also malate (not malic acid!; cf. 3.1.7) accumulation has been shown in nongreen tissues under conditions of high external cation (K^+, Na^+) concentrations (see Osmond, 1976b). However, diurnal fluctuations of malate levels corresponding to the oscillations of malic acid during CAM do not occur in nongreen tissues because the photosynthetic conversion of malate into carbohydrate (see Chap. 3) is lacking.

In stem succulents such as cacti (*Opuntia*, for example), only the outer cortical tissues rich in chloroplasts contribute substantially to CAM. In contrast, the pith tissues, which nearly lack chloroplasts, participate only to a very small extent (see Chap. 3).

Hence, it is reasonable to assume that the chloroplast-deficient pith tissue of all succulents, and also the colorless water-storing core in leaves, will lack CAM completely. As indicated above (Chap. 2.3.1) Kluge and Knapp (unpublished) found that the latter is true in *Aloe arborescens*.

On the other hand, the presence of a malate-synthesizing pathway together with the photosynthetic apparatus does not necessarily imply the occurrence of CAM. This is clearly shown in green leaves of C_4 plants which fix CO_2 from the atmosphere during photosynthesis nearly exclusively via β-carboxylation. In spite of a large malate-synthesizing capacity, CAM is lacking, perhaps due in part to an insufficient storage capacity for malic acid. An even more instructive example has been provided by McLaren and Thomas (1967), who compared the capacity for CAM in colorless and green tissue cultures isolated from the CAM plant *Kalanchoe crenata*. Analogous to what can be observed in roots, the colorless tissues were only able to perform β-carboxylation and malate synthesis (i.e., the "C_4 pathway of CO_2 fixation," cf. the above definitions). The green tissue showed both β-carboxylation as well as photosynthetic CO_2 fixation. However, in spite of the total biochemical machinery necessary for the occurrence of CAM in the cells of the green tissue cultures, no malic acid oscillation, i.e., no CAM, could be observed. Keeping in mind that cultured tissue cells mainly represent populations of meristematic cells with small vacuoles, it can be concluded that in this case the capacity to store malic acid is not sufficient to allow the performance of CAM.

From these considerations, the following generalization on the anatomical and cytological preconditions might be drawn: synthesis of malic acid via β-carboxylation of PEP, which is probably a feature of nearly all living plant cells, must be accompanied by the photosynthetic conversion of malic acid into carbohydrate. If the assimilating cells have also large storage vacuoles, CAM metabolism may be expected.

2.5 The Architecture and Ultrastructure of CAM-Performing Cells

2.5.1 Light Microscope Observations

The typical CAM photosynthetic cell is a large, thin-walled structure with a narrow peripheral band of cytoplasm. In the light microscope, these cells appear as empty spheres primarily because of the huge water-storing vacuoles which dominate (Fig. 2.3a). The number of chloroplasts per cell appears small in comparison with nonsucculent photosynthetic cells, but this may be illusionary because of the large cell size.

Typically, these tissues have large intercellular air spaces, perhaps facilitating gas diffusion within the voluminous water-storing tissues.

2.5.2 Electron Microscope Observations

Electron micrographs of CAM cells show the same nearly empty cells because of the dominating vacuoles. It can be seen (Fig. 2.5) that the peripheral cytoplasm surrounding the vacuole is extremely small, giving the appearance of little space for organelles such as chloroplasts or mitochondria. The electron micrographs of

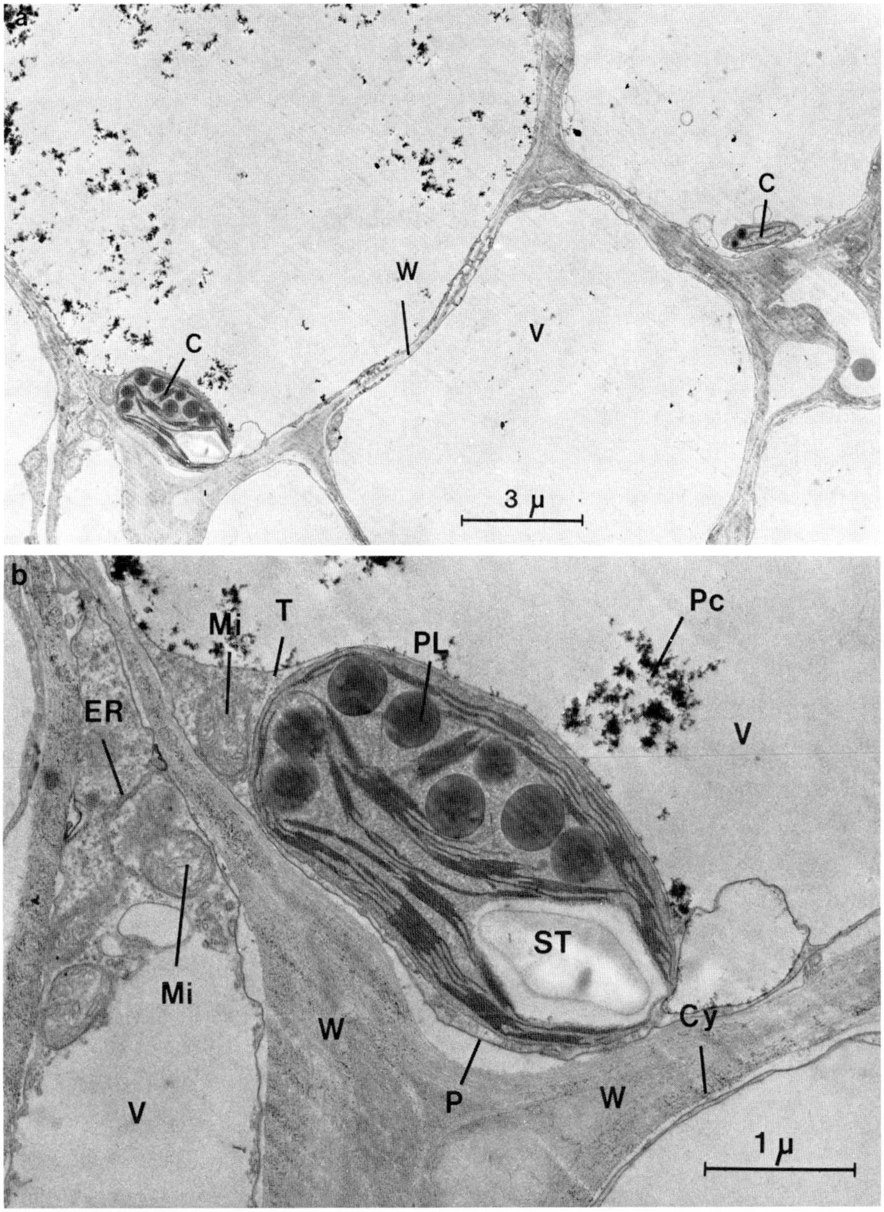

Fig. 2.5a and b. *Bryophyllum pinnatum*. **a** Section through CAM-performing mesophyll cells. **b** Detail of **a** showing a chloroplast (*C*) of the granal type with plastoglobuli (*PL*) and a starch granule (*ST*). Note the extremely thin layer of cytoplasm (*CY*) along the cell wall. *W*, cell wall; *Mi*, mitochondrion; *ER*, endoplasmic reticulum; *T*, tonoplast; *P*, plasmalemma, *V*, vacuole, *Pc*, precipitated content of vacuole. (electron micrograph courtesy of Dr. W. W. Thomson)

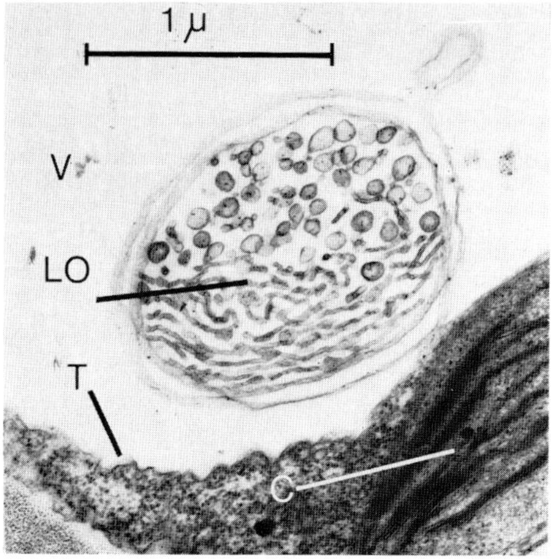

Fig. 2.6. Lomasome-like vesicles in the cells of *Mesembryanthemum crystallinum* in the CAM stage. *V*=Vacuole, *LO*=Lomasome, *T*=Tonoplast, *C*=Chloroplast (from Kramer and von Willert, 1972)

CAM cell vacuoles often appear filled with precipitations, probably due to tannins or tannin-like substances (see Fig. 2.5). Electron micrographs of von Willert and Kramer (1972) suggest membranous compartmentation within the vacuoles of *Mesembryanthemum crystallinum*. Also bladder-like vesicles extending from the cytoplasm into the vacuole were seen in the mesophyll cells of the same species. These vesicles consist of an envelope (double membrane) which encloses rather densely packed tubuli (Fig. 2.6) resembling lomasomes and interpreted as such by von Willert and Kramer. At the moment, there is no clear suggestion if these lomasome-like vesicles fulfill a specific role in CAM of *Mesembryanthemum crystallinum*.

By far most electron microscope studies of CAM have emphasized the ultrastructure of chloroplasts. Because of the similar CO_2-fixation biochemical pathways of CAM and C_4 photosynthesis plants, the ultrastructure of both has been compared. Early investigations of C_4 plant chloroplasts showed a peripheral reticulum, i.e., a layer of tubuli localized just below the chloroplast envelope (Laetsch, 1968, 1974). It was argued that the peripheral reticulum might be a typical feature of C_4 chloroplasts. Considering the biochemical similarities of the C_4 and CAM-type photosynthesis, the occurrence of chloroplasts having a peripheral reticulum was also predicted for CAM plants. However, there is now overwhelming evidence that the reticulum type of chloroplast is not a typical feature of C_4 photosynthesis plants or for plant species having CAM. The chloroplasts of CAM plants show no peculiarities unique and essential for the CAM metabolic pathway. Both the granal- (von Willert and Kramer, 1972; see Fig. 2.5), and agranal-type chloroplast (see Fig. 2.7) can be found in CAM plants. CAM chloroplasts may be remarkably rich in starch (see Fig. 2.8) consistent with the concept that the CO_2 acceptor (PEP) in β-carboxylation during dark CO_2 fixation is from the photosynthetically filled starch or glucan pool (cf. Chap. 3.1.5). In *Echinocactus acanthodes*, the chloroplasts of the outer cortical tissue contain

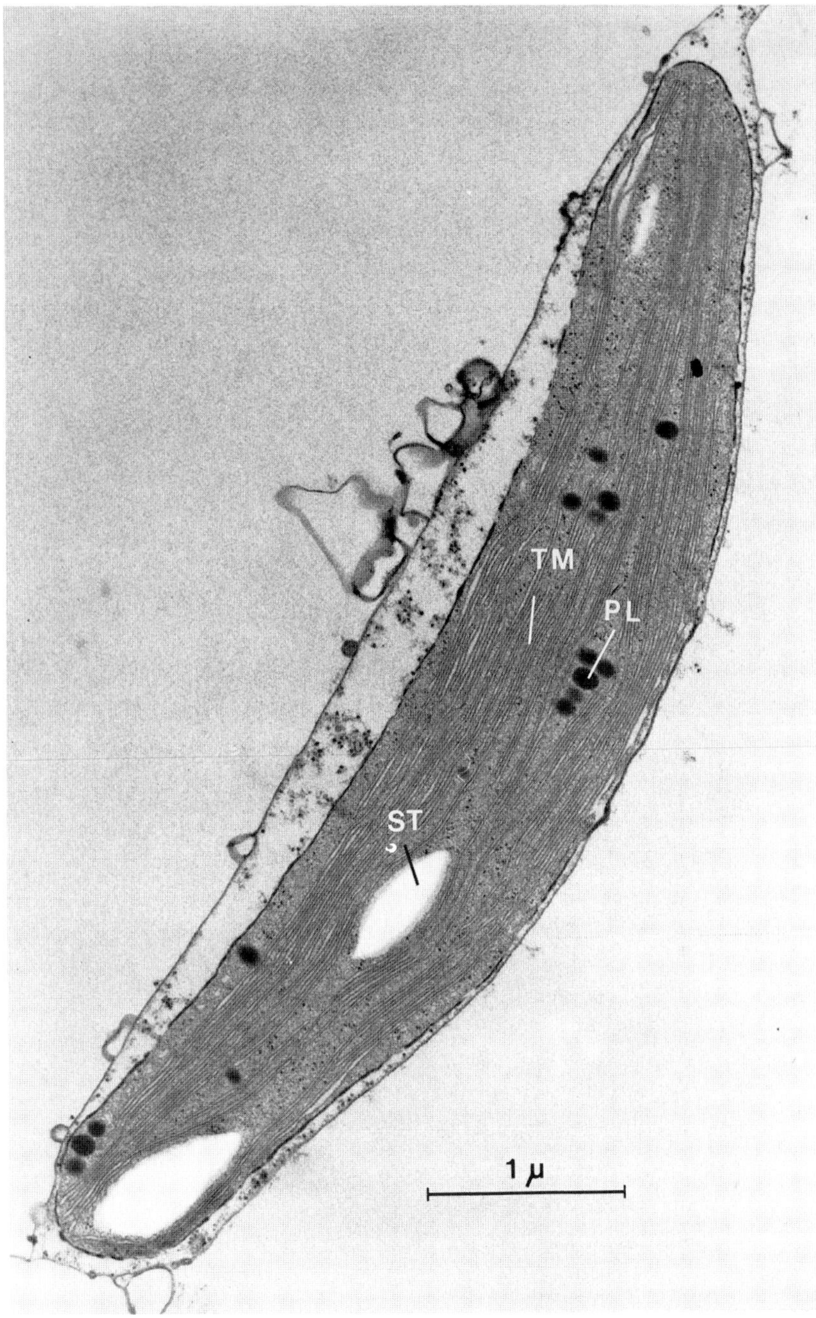

Fig. 2.7. Agranal chloroplast from the mesophyll of *Cotyledon salmiana*. *ST*, Starch granule; *PL*, Plastoglobulus; *TM*, Thylacoid membranes (electron micrograph by courtesey of Dr. Avadhani and Dr. Laetsch)

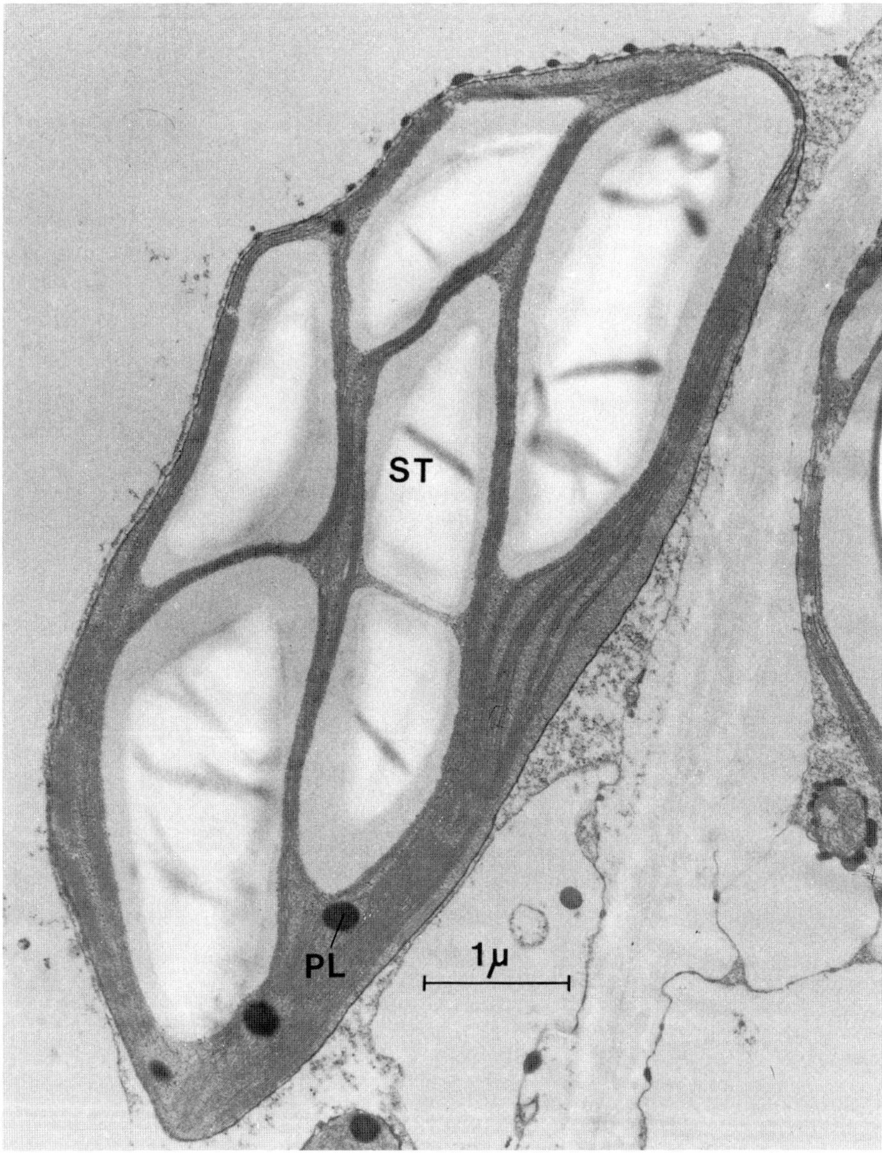

Fig. 2.8. Chloroplast from the mesophyll of *Kalanchoe daigremontiana*. The chloroplast has many large starch grains (*ST*) typical for the situation at the end of light period. *PL*, plastoglobulus (electron micrograph by courtesy of Dr. Avadhani and Dr. Laetsch)

numerous osmophilic plastoglobuli (Fig. 2.9), whereas the chloroplasts of the inner cortex bordering the pith tissue have both starch and plastoglobuli (Thomson and Platt, 1973). It was assumed that in *Echinocactus*, storage lipid may serve as a carbon reserve. It seems possible that the lipids stored in the plastoglobuli might be linked via acetyl-CoA to the starch pool. Also in *Bryophyllum pinnatum* (Crassulaceae) plastoglobuli can be observed in the chloroplasts (Fig. 2.5). However, it

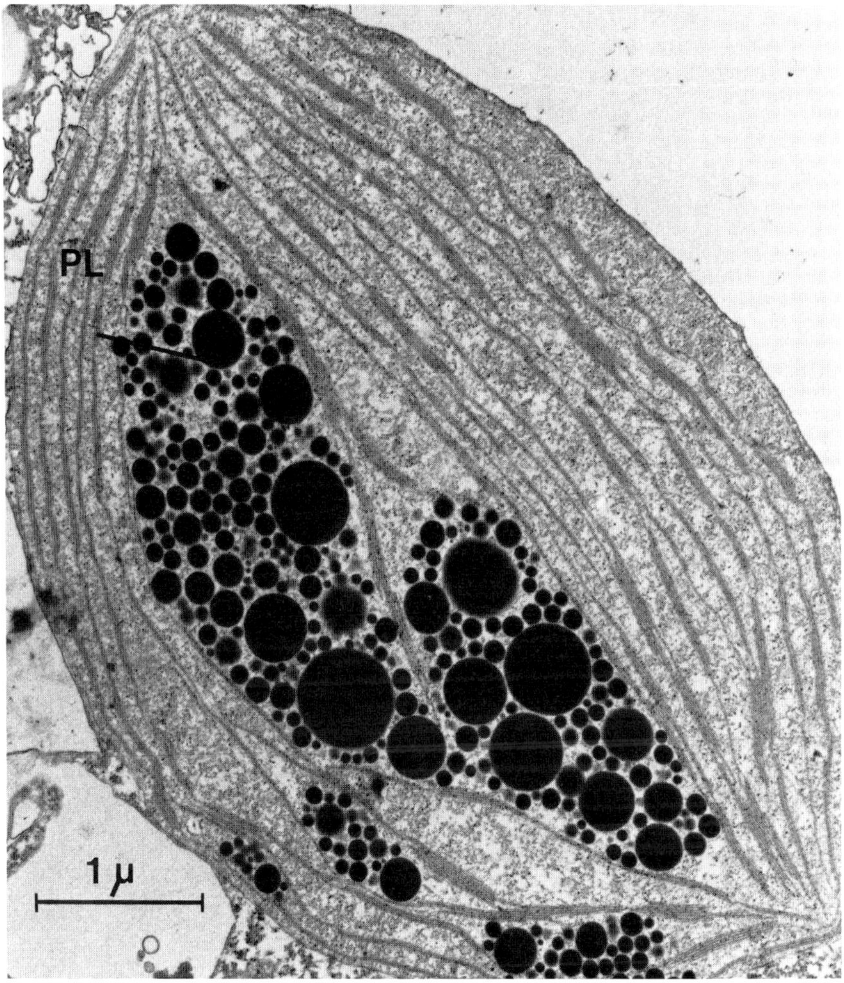

Fig. 2.9. Chloroplast from the CAM-performing mesophyll of *Echinocactus acanthodes* showing an extremely high density of plastoglobuli (*PL*) (by courtesy of Dr. W.W.Thomson)

still remains to be established if this phenomenon is a widespread feature among CAM plants; and if the lipid fluctuates diurnally similar to the total glucan pool. At least in *Echinocactus* it was observed by Kluge and Saupe (unpublished) in some preliminary investigations that carbon converted via CAM did not enter the lipid fraction, and that the number and size of the plastoglobuli did not change during the diurnal cycle of CAM. This would suggest that the lipids are not directly involved in carbon assimilation by CAM.

In chloroplasts of *Kalanchoe pinata*, paracrystalline protein inclusions have been described (Lee and Thompson, 1973). These inclusions, designated as stromacenters, consist of hexagonal subunits and occur in the plastids of the mesophyll surrounding the vascular tissues. Stromacenters and similar protein aggregates are also reported to occur in C_3 plants (Coleoptile of *Avena; Phajus walli-*

chii, Phaseolus vulgaris; see Lee and Thompson, 1973), and reports on stromacenters in CAM chloroplasts other than of *Kalanchoe* are lacking. Hence, there seem to be no striking arguments in favor of a specific role of stromacenters in CAM as suggested by Lee and Thompson (1973) and Thompson et al. (1977).

Kramer and von Willert (1972) observed in *Kalanchoe daigremontiana, Kalanchoe tubiflora,* and *Mesembryanthemum crystallinum* that in the "deacidified" phase of CAM, when the malic acid content of the cells is low, the chloroplasts and surrounding cytoplasm were attached to the cell wall. In contrast, during the "acidified state" (i.e., high level of malic acid) the chloroplasts were detached from the wall, leaving a space between the plasmalemma and the wall. This space has been shown to be filled by fibrillous structures resembling the Hecht'sche Fäden in plasmolyzed cells. Kramer and von Willert (1972) interpreted the space between plasmalemma as a "vacuole-like" compartment where malic acid could be stored. However, it is possible that the above phenomenon is a systematical artifact caused by the fixation of the tissue during its preparation for electron microscopy. Because it occurs only in the acidified state, it may reflect changing sensitivity of the cytoplasm to the fixing agents because of the acid diurnal rhythm of the CAM-performing cells. Also, during the accumulation of malic acid, the effective osmotic concentration of the cells changes substantially. Thus a careful adjustment of the osmotic concentrations of the fixation medium is necessary to maintain constant iso-osmotic conditions during the preparation of the tissue. Unfortunately, such precautions are frequently overlooked in the study of the fine structure of CAM plants.

As discussed later (see Chap. 3.2.4) in detail, CAM plants perform photorespiration as do C_3 plants. Hence, in this connection it is important to note that microbodies as a structural precondition of photorespiration have been demonstrated in the photosynthesizing cells of *Kalanchoe daigremontiana* (Kapil et al., 1975).

Kapil et al. (1975) described also unusual "microcylinders" having been observed in some individuals of *K. daigremontiana.* These structures "consist of a hollow cylinder 90–160 nm in diameter and up to 2 μm or more in length, around which 18–20 or more minute tubules are wound in a steep helix. The tubules are only ca. 9 nm in diameter, hence, are much smaller than conventional microtubules." Kapil and his co-workers suggest that the above structures might be due to a virus infection. However, they argue that in any case the "microcylinders" may prove to be of considerable importance for investigators of CAM.

Summarizing the known features of the ultrastructure of CAM plants, with the exception of the very large vacuoles, there are no known structures which appear unique for the occurrence of the CAM pathway. The cytological characteristics of CAM-performing cells are observed also in photosynthesizing tissues of other angiosperms. We conclude, therefore, that the feature of CAM is the presence of the proper enzymatic complement in chloroplast-containing cells which have immense central vacuoles capable of storing large quantities of aqueous solutions of malic acid.

3. The Metabolic Pathway of CAM

CAM is characterized by two segmental major metabolic sequences separated in time, one occurring in the night and the other during the day. The considerations of this chapter follow this sequential scheme. Hence, first the reactions dominating at night will be discussed, followed by a consideration of the light-dependent parts of CAM. Finally, a picture of total carbon flow of CAM will be drawn. The important problems of CAM regulation are covered in Chapter 4, which also includes the enzymology of CAM. Aspects of the CO_2 exchange, which are partially a direct consequence of the CAM metabolic pathway, will be considered in Chapter 5.

3.1 The Processes of the Dark Period

3.1.1 Early History

The first observation of the diurnal fluctuation of organic acids in succulent plants is rather difficult to assess. Osmond (1978) believes that M. Grew in 1682 was the first to realize acid accumulation in a succulent plant *(Aloe)*. As early as 1804, De Saussure noted that darkened *Opuntia* did not liberate CO_2, but did take up CO_2. Hence, De Saussure was in fact the first who observed dark CO_2 fixation in a CAM plant (see Chap. 5.1.1). The causal connection between dark CO_2 fixation and acid synthesis, however, was not realized until 1948 (Thurlow and Bonner, 1948; Thomas, 1949). In 1815, Heyne noted remarkable taste differences in diurnal alterations of the leaf acidity in *Bryophyllum calcycinum*. He wrote:

> "The leaves of the *Cotyledon calycina*, a plant called by Mr. Salisbury *Bryophyllum calycinum*, which on the whole have a herbaceous taste, are in the morning as acid as sorrel, if not more so; as the day advances, they lose their acidity, and are tasteless about noon and become bitterish towards evening."
>
> (Heyne, 1815; quoted after Wolf, 1960)

As quoted in Kraus (1884), shortly after Heyne also H. F. Link observed acid fluctuations in *Bryophyllum*. Then in 1865, J. Liebig recorded day-night reciprocal carbohydrate-acid fluctuations in succulents (see Kraus, 1884), and Mayer in 1887 noted that organic acids produced in the dark were converted to carbohydrates during the following light period. The early history of the development of the metabolic aspects of CAM has been recorded in reviews by Evans (1932), Bennet-Clark (1933a, b, 1937, 1949), Burris (1953), Davies (1959), Wolf (1960), Ranson and Thomas (1960), Walker (1962), Beevers et al. (1966), Ting (1971), Ting et al. (1972), and Black (1973), and in theses by Bradbeer (1954), Haidri (1955b), Walker (1956a), Brown (1956), Avadhani (1957), Lyndon (1957), Stiller (1959), Champigny (1960), Khan (1969), Rouhani (1972), Sutton (1974a), and Szarek (1974).

Extensive research by Richards (1915), Bennet-Clark (1933a, b), Wolf (1937), and then by such investigators as Pucher, Vickery, Thomas, Beevers, and Ranson firmly established the relationship between dark CO_2 fixation and organic acid synthesis in succulent plants. Furthermore, during this early period before 1950 it was clear that the organic acids which accumulated during the dark phase, termed "acidification" by the early workers, were nearly quantitatively converted to carbohydrates (viz., starch) during the subsequent light or deacidification period (see Bennet-Clark, 1933 for example).

3.1.2 Dark CO_2 Fixation and Its First Product

It is now generally accepted that dark fixation of CO_2 is the key reaction in CAM. Virtually all experiments conducted to date substantiate the hypothesis that malate[1] is the first and primary stable product of CO_2 fixation in CAM. Thurlow and Bonner and Bonner and Bonner in 1948 were perhaps the first to show dark $^{14}CO_2$ fixation into malate as well as other organic acids including citrate, isocitrate, fumarate, and succinate. Yet the equilibration of malate-^{14}C with the other organic acids was slow. Subsequently Thomas and Ranson (1954) reported the stoichiometry between CO_2 fixation and acid synthesis.

Vickery's early experiments (1954a, b) indicated that during the early phase of acidification, there was more malate produced than starch consumed. Hence, he inferred CO_2 uptake. Finally, based on data of Beevers and Gibbs (1954) and Stiller (1959), suggesting both glycolytic and pentose phosphate activity in supplying substrate from starch for malate synthesis, Ranson and Thomas (1960) proposed the following hypothesis for malate synthesis:

$$C_6H_{12}O_6 + 6O_2 \rightarrow 6H_2O + 6CO_2 \quad (3.1)$$

$$C_6H_{12}O_6 + 2CO_2 \rightarrow 2C_4H_6O_5 \quad (3.2)$$

where the degree of malate synthesis depended on the balance between Eqs. (3.1) and (3.2). The hexose moiety comes from glucan and the CO_2 in Eq. (3.2) can result either exogenously or from Eq. (3.1).

Saltman et al. (1957), using the eloquent techniques developed by Calvin's research group at Berkeley, showed that after 1 min of dark $^{14}CO_2$ fixation,

[1] It is a characteristic of CAM that malic acid is stored. However, most likely within the cytoplasm it is ionized as malate. Therefore, we use the term "malate" when we are talking about everything except the storage in the vacuoles.

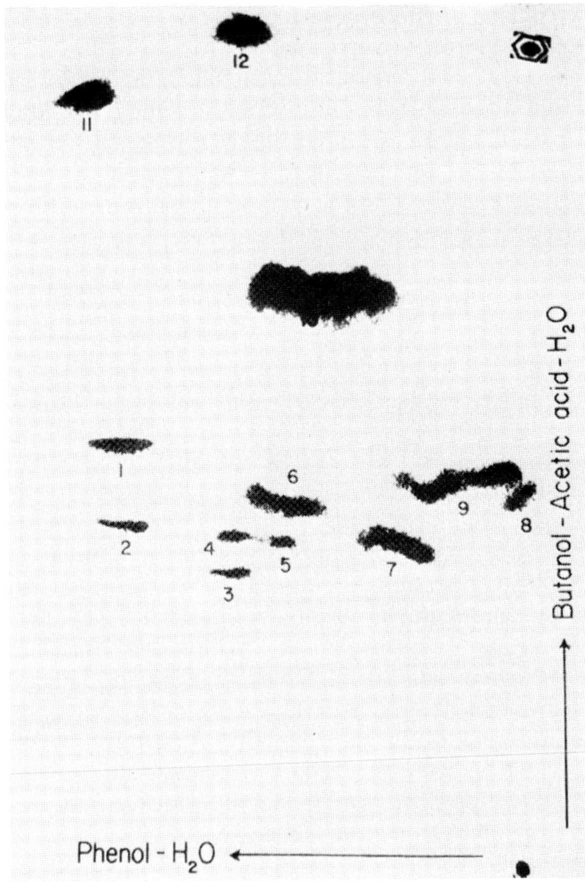

Fig. 3.1. Autoradiogram of two dimensional paper chromatogram of 60 minute dark $^{14}CO_2$ fixation by *Bryophyllum calycinum*. *1*, alanine; *2*, glutamine; *3*, asparagine; *4*, glycine; *5*, serine; *6*, glutamate; *7*, aspartate; *8*, citrate; *9*, isocitrate; *10*, malate; *11*, fumarate; *12*, succinate. The main labeled compound is malate. Other metabolically related organic and amino acids also appear. The only products after a 6-s exposure are malic and aspartic acids (data of P. Saltman et al., Plant Physiol. *32*, 197–200, 1957, by permission)

malate had 60% of the ^{14}C, and furthermore, even after 60 min no carbohydrates or phosphorylated compounds were labeled (Fig. 3.1). Since the first stable products detectable by chromatography were malate and aspartate, they deduced that oxalacetate was the first intermediate. The following two reactions were proposed:

$$\text{Pyruvate} + CO_2 \rightarrow \text{oxalacetate} \qquad (3.3)$$

$$\text{P-enolpyruvate} + CO_2 \rightarrow \text{oxalacetate}\,. \qquad (3.4)$$

Subsequently, additional research by Walker (1957) and others suggested that P-enolpyruvate was the main substrate for carboxylation rather than pyruvate, and furthermore, the enzyme, P-enolpyruvate carboxylase, discovered by Bandurski and Greiner (1953), was strongly implicated.

It was immediately recognized that the reaction catalyzed by malate dehydrogenase (Walker, 1957) probably was responsible for the reduction of oxalacetate to malate and aspartate aminotransaminase was responsible for aspartate formation.

It is now generally accepted that the metabolic pathway for the majority of the malate and aspartate synthesis in succulent plants is:

$$PEP + CO_2 \rightarrow OAA \begin{smallmatrix} \nearrow \text{malate} \\ \searrow \text{aspartate} . \end{smallmatrix} \tag{3.5}$$

3.1.3 Secondary Products and Organic Acids Other Than Malic

Detailed kinetics of dark CO_2 fixation product formation have not been reported. In *Bryophyllum calycinum*, after 1 min of dark CO_2 fixation, 60% of the label is in malate and 2% in aspartate. Already 11% was in isocitrate, and 4% in citrate and glutamate, and trace amounts in glutamine (Saltman et al., 1957). By 2 h of continuous $^{14}CO_2$ fixation, ^{14}C appeared in alanine, succinate, fumarate, serine, glycine, and asparagine. In *Opuntia*, a 2-h dark $^{14}CO_2$ period resulted in 66% malate, 8.5% aspartate-glutamate, 4% alanine, 10% succinate, 6% citrate, 4% isocitrate, 1.4% glycine, and 0.5% fumarate (Ting and Dugger, 1968). These data clearly indicate that malate, aspartate, citrate, isocitrate, succinate, glutamate, alanine, and glycine are the most important products of dark $^{14}CO_2$ fixation. Although difficult to determine accurately, oxalacetate, α-ketoglutarate, and pyruvate are prominent products, and in fact all evidence indicates that oxalacetate is the first product of CO_2 fixation (Table 3.1).

Since virtually all of the acids of the citric acid cycle were labeled, the question was asked early to what extent the mitochondrial citric acid cycle was involved. In this regard, Brown (cited in Ranson and Thomas, 1960) showed continued malate synthesis by *Bryophyllum* when in the presence of malonate, an inhibitor decreasing oxygen consumption and blocking the continued cycle.

Table 3.1. Primary products of dark CO_2 fixation[a]

Organic acids	Amino acids
Malate	Aspartate
Oxalacetate	Glutamate
Citrate	Alanine
Pyruvate	Glycine
Isocitrate	Glutamine
Fumarate	Asparagine
α-ketoglutarate	
Succinate	
Oxalate	

[a] From a variety of literature sources.

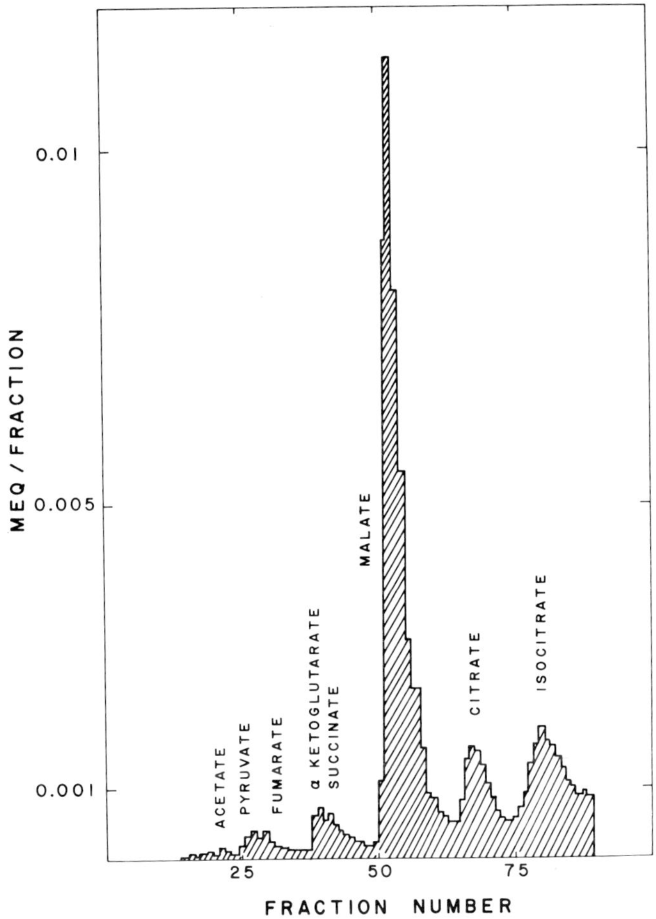

Fig. 3.2. Separation of *Kalanchoe blossfeldiana* organic acids by elution from a silicic acid column. Malic acid is the most abundant organic acid (data of O. Queiroz, by permission)

The evidence is now overwhelming that malic acid accounts for the diurnal fluctuation of organic acids during CAM. Vickery in a series of comprehensive investigations (Vickery, 1952, 1953, 1954a, 1957; Pucher et al., 1949) studied the diurnal variation of citric, isocitric, and malic acids under a variety of conditions of light and temperature and showed that, although isocitric acid may be the predominant organic acid, by and large it is malic acid that accounts for the 100–200 mEq per kg fresh weight diurnal fluctuation. Fluctuations of citric and isocitric acids are minimal.

Figure 3.2 shows a typical silicic acid chromatogram of organic acids from a crassulacean succulent. In this case, with *Kalanchoe blossfeldiana*, malic acid is the dominant acid.

Nordal and his associates have studied organic acids in a variety of succulent CAM plants. Phorbic acid, an 8-carbon tricarboxylic acid, was detected in *Euphorbia sp.*, *Cereus*, *Opuntia*, *Kleinia*, and *Portulaca* (Nordal and Ogner, 1964;

Bernatek et al., 1963). In *Agave americana*, they identified piscidic acid (p-hydroxy-benzyl-tartaric acid, a dicarboxylic aromatic acid) which is also known in the Leguminosae and Liliaceae (Nordal and Ogner, 1964). They estimated that piscidic acid composed approximately half of the accumulated organic acids in *Agave*. Since phorbic acid was a major component of the total organic acids in *Euphorbia resinifera*, they suggested that these two acids, phorbic and piscidic, may play an important role in CAM.

Ting (unpublished) was able to show, however, that the main organic acids produced by dark CO_2 fixation in *Euphorbia loricata* did not cochromatograph with pure, authentic phorbic acid (generously supplied by Dr. Nordal) and that the main organic acid was, in fact, malic acid. Hence, it is unlikely that phorbic and piscidic acid play a predominant role in CAM.

Finally, Nordal and Benson (1969) found phorbic acid in the latex fraction after light $^{14}CO_2$ fixation by *Euphorbia resinifera*. They concluded on the basis of their experiments that phorbic acid is a secondary product produced in the latiferous vessels and is not a direct photosynthetic or dark fixation product.

3.1.4 The Active Chemical Species of "CO_2"

Current information concerning the active species of CO_2 (i.e., CO_2, HCO_3^-, or CO_3^{--}) for the PEP carboxylase reaction is somewhat equivocal. Maruyama et al. (1966) concluded that bicarbonate, and not CO_2, was the substrate for carboxylation. In their experiments using uniformly labeled $HC^{18}O_3^-$, they assumed a nucleophilic attack of one ^{18}O of the $HC^{18}O_3^-$ on the phosphorus of P-enolpyruvate. This would result in two ^{18}O atoms occurring in the β-carboxyl group of oxalacetate. The remaining ^{18}O atom would be in the liberated inorganic phosphate. They showed that there was slightly more ^{18}O in oxalacetate than the theory predicted. Hence, they concluded that HCO_3^-, and not CO_2, is the active species. Their experiments may be somewhat suspect, however, because of their assumption that OH^- at the active site was in equilibration with the surrounding solvent (Cooper et al., 1968).

Cooper and Wood (1971), using a kinetic experiment which they developed, also concluded that HCO_2^- and not CO_2 is the active species. In their experiment with maize PEP carboxylase, they followed carboxylation kinetics with either added $^{14}CO_2$ or $H^{14}CO_3^-$. Here they showed that the reaction was faster with HCO_3^- than with CO_2, presumably because of the delay required for CO_2 to be converted to HCO_3^-:

$$CO_2 \rightleftarrows H_2CO_3 \rightleftarrows H^+ + HCO_3^- . \tag{3.6}$$

However, when CO_2 plus carbonic anhydrase were added together, they did not obtain a rate comparable to HCO_3^- alone, but 50% of the rate. Hence, a discrepancy was noted. In comparable experiments with other carboxylases including RudP carboxylase, the predicted data more nearly agreed with theory.

Waygood et al. (1969), however, using PEP carboxylase from maize leaves and the same procedure, obtained opposite results. They concluded, therefore, that maize PEP carboxylase uses CO_2 and not HCO_3^-.

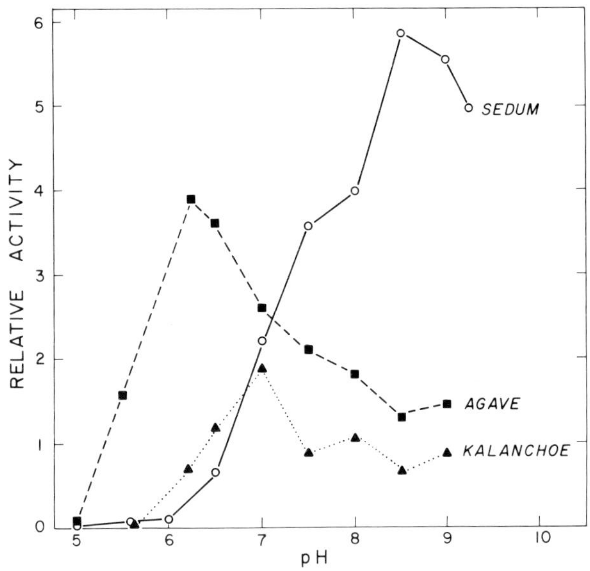

Fig. 3.3. pH optimum curves for PEP carboxylase isolated from three different CAM plants. The pH optimum is variable among the species tested, although most have high optima more consistent with HCO_3^- as the inorganic carbon substrate (after Kluge and Osmond, 1972)

More recently, Coombs et al. (1975) reinvestigated the inorganic carbon substrate of carboxylation and found by both isotopic and spectrophotometric methods that HCO_3^- and not CO_2 was the active species. They used PEP carboxylase isolated from *Pennisetum purpureum* and *Pisum sativum.*

Filmer and Cooper (1970) argued that carboxylases with high apparent Km's for total CO_2 (i.e., $CO_2 + HCO_3^-$) in the range of 5–10 mM were largely CO_2 enzymes, whereas those with low apparent Km's were HCO_3^- enzymes. In this regard most estimates of plant PEP carboxylase Km's for total CO_2 are in fact low. Our own estimates for three PEP carboxylases isoenzymes of cotton were 0.14, 0.05 and 0.06 mM (Mukerji and Ting, 1971). Estimates for PEP carboxylases from *Atriplex spongiosa* and *A. hastata* were about 0.02 mM (Ting and Osmond, 1973a, b). The Tate and Lyle research group found a Km (total CO_2) of 2 mM for *Pennisetum* PEP carboxylase in the absence of 5 mM glucose-6-phosphate, but a reduction to 0.8 mM in the presence of glucose-6-phosphate (Coombs et al., 1973).

To some extent pH optimum curves for the carboxylases give information about the active species of CO_2. With a pK for HCO_3^- of 8.4, the higher the pH, the less CO_2 available. By and large, plant PEP carboxylases seem to have rather high pH optimum. Kluge and Osmond (1972), however, using a variety of succulents, showed that the pH optimum was variable among species (Fig. 3.3).

Since multiple molecular forms of PEP carboxylase are known, there is no a priori reason to assume that all would use either CO_2 or HCO_3^-. Indeed Jacoby and Laties (1971) suggested that the different molecular forms may differ in the substrate form of CO_2.

By and large, however, the available data support the notion that HCO_3^- is the active species for carboxylation.

Other carboxylases which are known to function in CAM plants, such as RuDP carboxylase and PEP carboxykinase, use CO_2 as the active carboxylating species (Cooper et al., 1968, 1969).

3.1.5 Generation of P-Enolpyruvate, the CO_2 Acceptor in Dark CO_2 Fixation

The many experiments by Vickery and others have clearly established the reciprocal relationship between malic acid and stored carbohydrate. The latter is primarily starch or other polyglucose (glucan). During acidification, as glucan is depleted, malic acid accumulates. Furthermore, it was demonstrated (Kluge et al., 1975) by pulse chase experiments that ^{14}C is transferred from glucan labeled in photosynthesis to malate via CAM during the following night.

Further, the level of glucan is reduced less if dark CO_2 fixation is inhibited by CO_2-free air or high temperatures. Finally, Cockburn and McAulay (1977) demonstrated that the levels of PEP and other intermediates of the glycolytic pathway increased following transition from normal to CO_2-free air in the dark.

Hence, the notion that the carbon for malic acid synthesis comes from glucan and free CO_2 is firmly established. It is, therefore assumed that:

$$\text{Glucan} \rightarrow \text{P-enolpyruvate}. \tag{3.7}$$

The exact metabolic pathway from storage carbohydrate to P-enolpyruvate is not clear. Kluge (1969a) with *Bryophyllum* showed that there was a reciprocal relationship between P-glycerate and ribulosephosphate with P-glycerate high in the dark. The cellular pool of P-enolpyruvate, similarly to P-glycerate, was higher in the dark than in the light, but levels of other phosphate esters such as fructose-6-phosphate, fructose-1,6-diphosphate, and glucose-6-phosphate were nearly constant day and night. These data suggest, therefore, that the availability of intermediates for P-enolpyruvate synthesis remains high or even increases at night, and indeed, P-enolpyruvate itself increases (see also Cockburn and McAulay, 1977).

The question of the glycolytic sequence versus the pentose phosphate pathway for P-enolpyruvate synthesis was first approached by Beevers and Gibbs (1954). They estimated, by following ^{14}C label in glucose-1-^{14}C and glucose-6-^{14}C, a ratio of 54% C-6/C-1 in *Bryophyllum* leaves, suggesting approximately equal participation of the two pathways. Further, Stiller (1959) in her Ph. D. thesis reported extensive pentose phosphate cycle activity in *Kalanchoe*, and Khan (1969) at Lucknow University studied enzymes of the pentose phosphate pathway in the cactus, *Nopalea dejecta*. Sutton (1974a, b; 1975a, b, c) thoroughly investigated the glycolytic sequence in crassulacean succulents, and found sufficient enzyme activity to account for P-enolpyruvate synthesis during acidification. Sutton found that a decrease in a polyglucose glucan, but not starch, was sufficient to account for the malic acid accumulation. Further, the turnover and specific activity of free sugar were insufficient to implicate them in the pathway toward PEP synthesis. Low levels of glucose-6-phosphate dehydrogenase and the good stoichiometry between glucan decrease and malate increase led him to conclude that

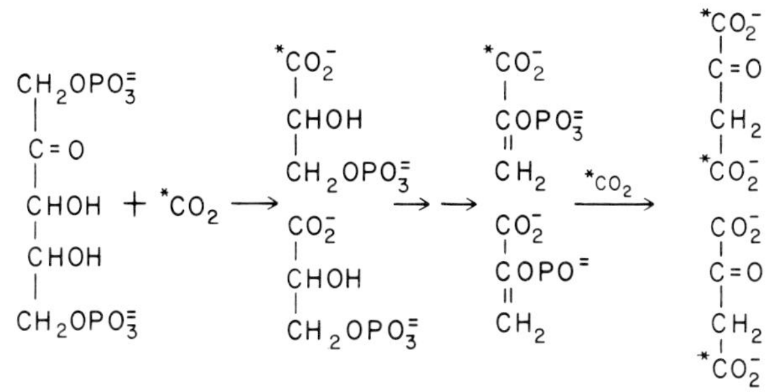

Fig. 3.4. The double carboxylation hypothesis of Bradbeer et al. (1958). The first carboxylation through RudP carboxylase results in one labeled PGA in the C-1 position and one unlabeled. After conversion to PEP, a second carboxylation through PEP carboxylase yields equal ^{14}C label in the C-4 of malate giving the ratios of 2 C-4 label to one C-1 label so frequently found in CAM plants

glycolysis was the source of PEP in CAM. Since α-amylase was not detectable, but phosphorylase was, the following pathway is assumed:

$$\text{glucan} + \text{iP} \rightarrow \text{glucose} - \text{P} \rightarrow \text{PEP}. \qquad (3.8)$$

More recently, however, using improved methods, Vieweg and Fekete (1977) were able to demonstrate α-amylase activity in *Kalanchoe daigremontiana* sufficiently high to account for substantial breakdown of starch. Hence, most likely both phosphorylase and α-amylase participate in starch mobilization.

It should be clear that as well as stored carbohydrate degradation to form P-enolpyruvate for malate synthesis in CAM, oxidative metabolism is proceeding concurrently and acetate must be generated for the citric acid cycle. However, the rate of malate synthesis may be five times greater than respiration. The enzymology of the citric acid cycle was studied in detail by Khan (1969).

A further complication and interesting feature, concerns participation of at least parts of the reductive pentose pathway (Calvin cycle). CAM, as presently understood, only occurs in chloroplast-containing cells which have a full reductive pentose phosphate cycle (see Chap. 2.4). A variety of types of experiments have implicated a portion of the photosynthetic pathway.

In their $^{14}CO_2$ experiment with *Bryophyllum*, Varner and Burrell (1950) recorded that there was an asymmetric labeling pattern between the one and four carboxyls of malate. After 2.5 h at 10° C, 54% of the ^{14}C was in 4-COOH and 30% in 1-COOH. Subsequently, Bradbeer et al. (1958) studied $^{14}CO_2$ fixation by both *Kalanchoe* and *Bryophyllum*, and using enzymic and chemical techniques, observed that the distribution between 4-COOH and 1-COOH was always about 2:1 and virtually no ^{14}C appeared in the middle carbons. Hence, they concluded that there was a precarboxylation of ribulose-1,5-diphosphate, resulting in an exact 2:1 labeling pattern (Fig. 3.4). Subsequently, many others confirmed the 2:1 labeling pattern in a variety of tissues.

One prediction of this hypothesis is that P-glycerate, the product of carboxylation of ribulosediphosphate, would appear as a labeled intermediate. In fact in

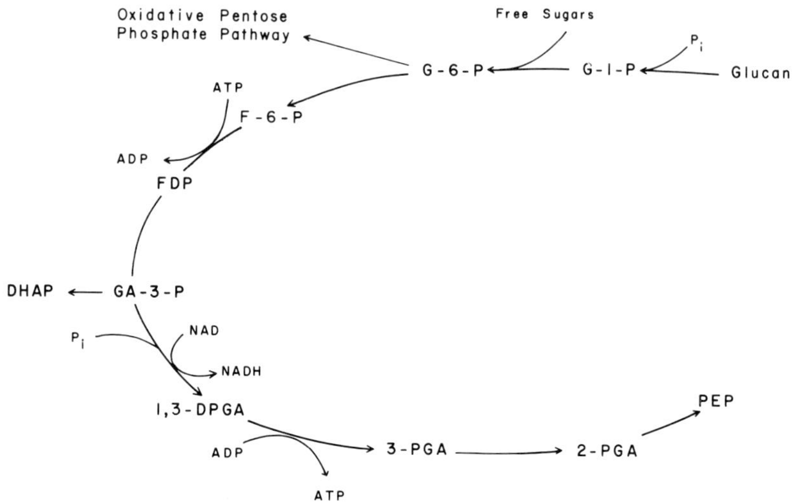

Fig. 3.5. The proposed carbon flow pathway from storage carbohydrates (glucan) through glycolysis to PEP. This is the current hypothesis stating that PEP is generated from storage carbohydrate at night through the reactions of the glycolytic sequence (redrawn from Sutton, 1974a)

one series of experiments, P-glycerate-^{14}C was found, but the results could not be repeated (Stiller, 1959). Black, however, has recently reported finding some ^{14}C labeled P-glycerate during dark $^{14}CO_2$ fixation by isolated *Sedum* mesophyll cells (Black, 1973).

Previously, it was argued that asymmetrical labeling could result because of nonrandom equilibration of the malate carboxyl groups (malate$\rightleftarrows$fumarate). This could occur by a certain proportion of the malate being diverted to a storage pool (e.g., vacuole), and the remainder flowing through the citric acid cycle (Ting, 1971).

Recently, Sutton and Osmond (1972), studying *Bryophyllum* and other succulents, reported that kinetically, after 1 min virtually all the ^{14}C from $^{14}CO_2$ was in the 4-COOH of malate; and after several hours, there was a slow randomization with 1-COOH toward an equal distribution. The usual malate degradation technique is to obtain crude malate enzyme from freeze-dried *Lactobacillus* (^{1}COO-$CHOH$-CH_2-$^{4}COO^{-} \rightarrow {}^{1}COO$-$COCH_3 + CO_2$). Sutton and Osmond, however, used an improved method of degradation with purified malate enzyme (Hatch, 1972) attempting to eliminate the possibility of randomization during the assay because of enzymes other than malate enzyme in the *Lactobacillus* preparation (e.g., fumarase).

Kluge et al. (1974, 1975) demonstrated that exogenously added P-enolpyruvate did not alter the ratio of label in C-4 to C-1, despite the fact that the added P-enolpyruvate increased carboxylation. Hence, there was no dilution of prelabeled P-enolpyruvate, as would be predicted by the double carboxylation hypothesis.

Cockburn and McAulay (1975a, b) in an eloquent experiment labeled malic acid with $^{13}CO_2$ during dark CO_2 fixation by *Kalanchoe*. Subsequent mass spectroscopic analysis showed ^{13}C in carbon-1 and carbon-4, but no double-labeled malic acid molecules (i.e., both in C-1 and C-4). They concluded that only a single carboxylation occurred (via PEP carboxylase) and randomization of carboxyl at the 4 position of malic acid with the 1 position accounted for the 2:1 distribution.

Hence, we are forced to conclude that there is a single carboxylation in CAM, and that the carboxyls of malic acid randomized through a reaction similar to fumarase. Figure 3.5 depicts the present hypothesis for carbon flow from carbohydrate to malic acid in CAM.

3.1.6 Depletion of Malate in the Dark

It is well known that during prolonged darkness CAM plants lose the malic acid which accumulated during the initial phase of the night (see Wolf, 1960). This decrease in malic acid level is accompanied by CO_2 release. Hence, there is little doubt that malate is the major source for CO_2 release in prolonged darkness. As shown by Kluge (1968a), even during a normal night some CO_2 may be released at the expense of malate.

It remains an open question whether or not malate is decarboxylated in the dark by malate enzyme or PEP carboxykinase (see Chap. 4.2.1).

The initial carboxylation reaction results in malate, which is known to turn over by decarboxylation at some rate less than carboxylation. There is the possibility of other less important carboxylations, e.g. the reaction given in Eq. (3.9).

$$\alpha\text{-ketoglutarate} + CO_2 - NADPH \rightleftarrows \text{isocitric} + NADP. \tag{3.9}$$

Since citric acid cycle intermediates are quickly labeled it is assumed that much subsequent metabolism of malate is by citric acid cycle reactions. Of course, the malate itself could enter the mitochondria, and no doubt does (Lips and Beevers, 1966a, b; MacLennan et al., 1963). Pyruvate and/or P-enolpyruvate resulting from decarboxylation probably enter the citric acid cycle via acetate. Bradbeer and Ranson (1963) found that ^{14}C-pyruvate supplied to *Kalanchoe* leaves in the dark was consumed by the citric acid cycle. Some was also found as free sugars. Hence, there is a slow equilibration of malate and malate degradation products with other metabolic events in the cell. Measurements of oxygen consumption during CAM metabolism can leave no doubt that the citric acid cycle is active in succulents (e.g., Szarek and Ting, 1974b; Kinraide and Behan, 1975; Kaplan et al., 1976a, b; 1977).

3.1.7 The Storage of Malic Acid

It is one of the most outstanding features of CAM that malate synthesized via dark CO_2 fixation is stored as malic acid and may accumulate up to 100–200 mEq kg^{-1} f.wt. It is most unlikely that this quantity of acid can be stored in the cytoplasm without a drastic decrease in cytoplasmic pH. The huge vacuoles of the CAM cells provide a convenient compartment where large quantities of aqueous malic acid may be reversibly stored. Kluge and Heininger (1973) showed indirectly the vacuolar storage pool by measuring the efflux kinetics of malic acid in leaf slices of *Kalanchoe daigremontiana*. Recently, Buser and Matile (1977) were able to isolate intact vacuoles from *K. daigremontiana* leaves and found on the average more than 90% of total malic acid of the leaves localized in the vacuoles. These data are in agreement with the values of Kluge and Heininger (1973).

It has been suggested that malic acid storage in the vacuoles might be integrated into the mechanism regulating the diurnal cycle of CAM (Wolf, 1960; Kluge and Heininger, 1973; von Willert and Kluge, 1973; Lüttge et al., 1975, 1977). There is evidence that malic acid transport into vacuoles is an active process, whereas the release back into the cytoplasm is passive (Lüttge and Ball, 1974; Lüttge et al., 1975; Denius and Homann, 1972). Although most evidence and observations indicate that light acts as the signal for malic acid release and decarboxylation, experimentally it can be shown that turgor created by malic acid accumulation might play a role (see Kluge and Lüttge, 1974; Lüttge and Ball, 1974; Lüttge et al., 1975, 1977; Chap. 4.2.3).

Light signals, in particular far red, may also trigger malic acid efflux from the vacuoles into the cytoplasm (Kluge, 1971b; Nalborczyk and La Croix, 1975).

However, both hypotheses are tentative.

3.2 The Processes of the Light Period

3.2.1 Deacidification and Malate Decarboxylation

After a period of dark CO_2 fixation resulting in the vacuolar accumulation of malic acid and decreased levels of stored carbohydrate, there is a rapid and marked decrease in stored malic acid when the plants experience light (Fig. 3.6). Malate, once mobilized for subsequent metabolic consumption, may inhibit P-enolpyruvate carboxylase (Ting, 1968) and reduce further carboxylation (Kluge, 1969; Queiroz, 1967; Ting and Osmond, 1973a, b). It is generally assumed that deacidification occurs because of malate decarboxylation (or oxalacetate decarboxylation) and concomitant release of CO_2.

Circumstantial and direct evidence is quite strong that NADP-dependent "malate enzyme" (E.C.1.1.1.40) participates in the decarboxylation and consumption of malate during deacidification (Walker, 1962) [see Chap. 4.2.1.4 and Eq. (4.6)]. Recently, Dittrich found also NAD-dependent "malate enzyme" in CAM plants (Dittrich, 1975). Dittrich et al. (1973) conducted an extensive survey of CAM plants for the presence of malate enzyme and P-enolpyruvate carboxykinase. They found that those CAM plants deficient in PEP carboxykinase were sufficient in malate enzyme; however, the reverse was not necessarily true. Those members of the Agavaceae, Aizoaceae, Asteraceae, Cactaceae, Crassulaceae, and Orchidaceae which were studied lacked PEP carboxykinase, but had sufficient malate enzyme to account for malate consumption. Asclepiadaceae, Bromeliaceae, Euphorbiaceae, and Vitaceae had high PEP carboxykinase, but also malate enzyme. In the Liliaceae, *Sansevieria* (two species) fell into the former group and *Aloe arborescens* and *Apiera spiralis* were in the latter group. Hence, there is a strong possibility that alternate pathways unique to different taxa may exist.

Studies by Milburn et al. (1968) indicated that in *Bryophyllum*, pyruvate was high during light deacidification, suggesting malate decarboxylation by malate

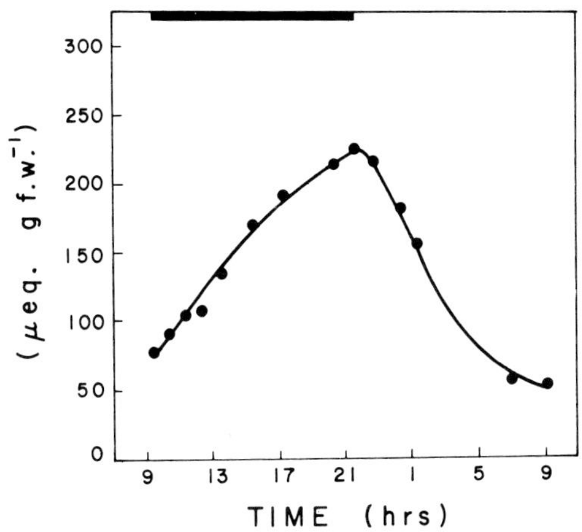

Fig. 3.6. Diurnal fluctuation of malic acid by a Crassulacean plant, *Kalanchoe tubiflora* (data of Sutton, 1974a)

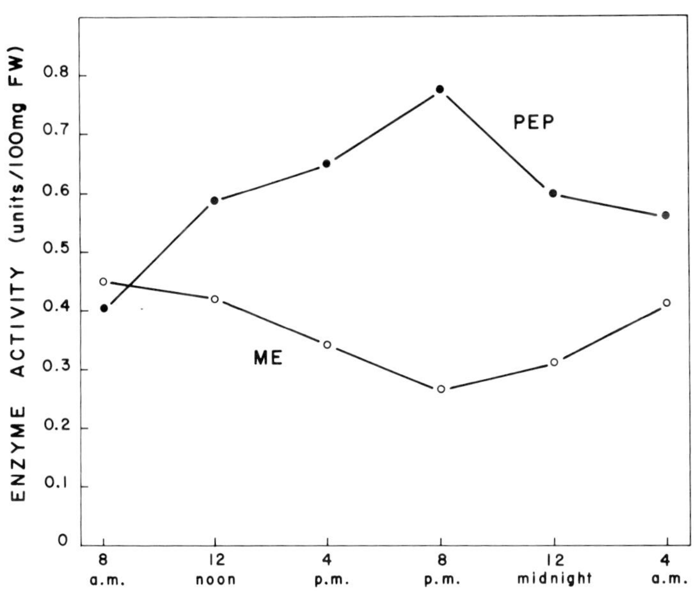

Fig. 3.7. The diurnal variation of PEP carboxylase (*PEP*) and malate enzyme (*ME*) in a cactus, *Nopalea dejecta*. PEP carboxylase activity is high during acidification and malate enzyme is high during deacidification. Assay run on preparations which were not desalted or otherwise treated (data of Mukerji, Ind. J. Biochem. Biophys. 5, 62–63, 1968, by permission)

enzyme. There is considerable evidence that both P-enolpyruvate carboxylase and malate enzyme fluctuate diurnally in the Crassulaceae and Cactaceae, and that the fluctuation is out of phase (Mukerji, 1968; Queiroz, 1968b, 1974; Khan, 1969). P-enolpyruvate carboxylase is high during the acidification portion of the diurnal cycle and malate enzyme is high during deacidification (Fig. 3.7). These data tend to implicate malate enzyme as the decarboxylating catalyst, at least in some.

3.2.2 The Fate of the Decarboxylation Products

3.2.2.1 Three-Carbon Fragments

Quantitatively the fate of pyruvate or other 3-carbon fragments in the light is far more important than in the dark (see Chap. 3.1.6). The critical question is whether the pyruvate is completely oxidized to CO_2 prior to carbon reassimilation or if it is quantitatively converted to carbohydrates, perhaps by glycolytic reversal. Early experiments clearly showed that malate was not converted directly to carbohydrate (Ranson and Thomas, 1960).

Champigny (1960), in a variety of different types of studies, suggested that malate may be degraded to CO_2 completely prior to carbon entering carbohydrates. In other studies, Haidri (1955a, b) infiltrated pyruvate-2-^{12}C into *Kalanchoe* leaves and from the distribution of carbon in hexoses concluded that pyruvate entered carbohydrates by a reversal of glycolysis, yet the data suggested that some pyruvate was oxidized to CO_2 and then incorporated into carbohydrates. Furthermore, Milburn et al. (1968) reported that in their experiment complete oxidation of the disappearing malate by the citric acid cycle would have required a 10-fold greater consumption of oxygen than they were able to measure. Despite these observations, there is a positive correlation between oxygen consumption in the light and total acidity (Fig. 3.8; Szarek and Ting, 1974b).

We must conclude from these experiments that: (1) malate is not converted to carbohydrate directly, (2) the 3-carbon fragment can be directly converted to carbohydrate, and (3) some of the 3-carbon fragment can be oxidized to CO_2, which then may be converted to carbohydrate by photosynthesis. It seems that most of the 3-carbon fragment is converted directly to carbohydrate by a reversal of glycolysis.

A complication of the latter hypothesis is the energy barrier which must be overcome to go from pyruvate to P-enolpyruvate prior to glycolytic reversal. One attractive hypothesis is that the light-activated pyruvate, Pi dikinase, converts pyruvate to P-enolpyruvate. The latter enzyme has been detected in succulent plants (Kluge and Osmond, 1971; Sugiyama and Laetsch, 1975). Since the ATP sufficient for the reaction probably results from photosynthetic phosphorylation, light regulation of carbohydrate synthesis from malate is indicated. Alternatively, in those succulents with high P-enolpyruvate carboxykinase, malate via oxaloacetate could be converted directly to P-enolpyruvate. Figure 3.9 outlines the possible metabolic fate of pyruvate in the light.

Energetically, starting with malate it would require about four ATP equivalents to attain the level of carbohydrate.

3.2.2.2 Carbon Dioxide

In the light malate decarboxylation, CO_2 is largely conserved by refixation through photosynthetic reactions (Kunitake and Saltman, 1958). In the latter investigation it was clearly shown that after a period of dark $^{14}CO_2$ fixation, if crassulacean succulents were transferred to light, much of the ^{14}C was retained. Most important was the fact that in the light labeled carbohydrates began to appear indicating that dark-fixed carbon was transferred to sugars.

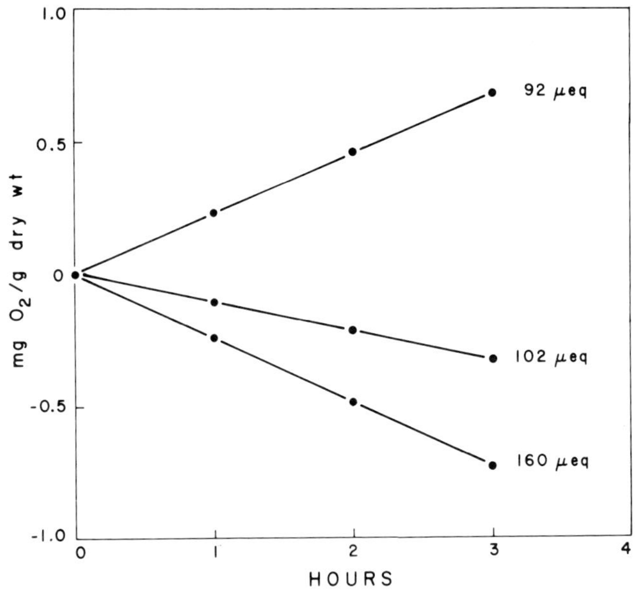

Fig. 3.8. Oxygen exchange by tissue sections of *Opuntia basilaris* as a function of tissue acidity, there is a trend toward oxygen consumption (data of S. Szarek and Ting, Plant Physiol. *54*, 829, 1974)

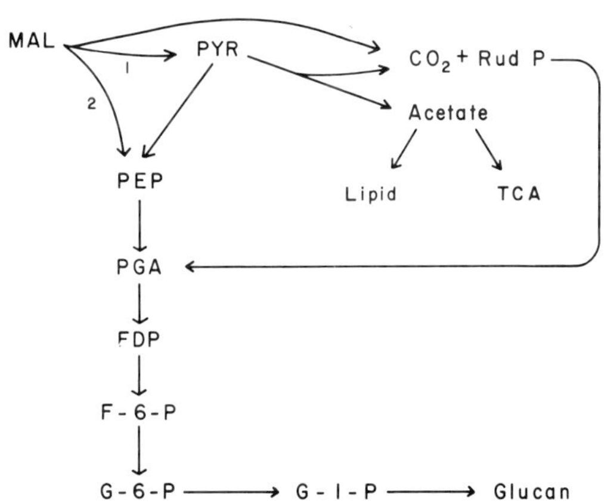

Fig. 3.9. Proposed metabolic flow of carbon by metabolism of pyruvate in the light. Most pyruvate carbon probably is returned to storage carbohydrate

In *Opuntia*, after a pulse of $^{14}CO_2$ supplied in the dark, $^{14}CO_2$ is released and labeled organic and amino acids decrease. If however, the pulse is followed by a light treatment, the release of $^{14}CO_2$ is decreased to about 10% of the dark release, and it can be shown by calculation that the retained CO_2 is quantitatively converted to photosynthetic products (Ting and Dugger, 1968).

Hence, in the light, and in particular when stomata are closed, CO_2 stored as malic acid is reutilized in photosynthetic reactions effectively conserving the dark assimilation of CO_2.

Table 3.2. Products of dark and light CO_2 fixation by *Opuntia*[a] (data of Ting and Dugger, 1968)

Product	Light[b]	Dark
% Lipid	0.22	0.0
% Water soluble	99.8	100.0
Amino acid	11.2	15.9
Carbohydrate	56.2	0.0
Organic acid	32.1	84.1
Citrate	7.4	8.7
Malate	61.7	74.3
Isocitrate	4.9	3.1
Succinate	14.7	9.4
Fumarate	2.5	0.3
Other	8.7	4.1

[a] Data expressed as % of each fraction after 2-h fixation period.
[b] Light $= 1.2 \times 10^5$ ergs cm^{-2} s^{-1}.

In other words, the malic acid synthesized during the night by dark CO_2 fixation can be interpreted in terms of CO_2 storage during the night. Because of this endogenous "CO_2 storage," photosynthesis in CAM plants can proceed with closed stomata. This peculiarity provides the ecological advantage of CAM as outlined in detail in Chapter 6.

3.2.3 Assimilation of Exogenous CO_2 in the Light

In CAM plants, little is known about photosynthesis using exogenous CO_2. In 1958 Champigny et al. reported data indicating to them that in the light, *Bryophyllum daigremontianum* assimilates $^{14}CO_2$ simultaneously by "photosynthetic carboxylation" i.e., ribulose diphosphate carboxylase, and "β-carboxylation," i.e., P-enolpyruvate carboxylase. Their experiments clearly indicated that malate was labeled in both dark and light, and that the extent of labeling in the light was greater than in the dark. Similar results were found with *Opuntia* (Ting and Dugger, 1968; Table 3.2). Avadhani et al. (1971) reported light fixation of $^{14}CO_2$ into phosphorylated compounds similar to typical photosynthesis and they further reported that the malate appearance and labeling was more similar to C_4- than to C_3-photosynthesis. Gibbs (1953) found that the carboxyls of malate became randomized faster in light than in dark.

Kluge (1969c, 1971b) also reported on changing label patterns after fixation of exogenous CO_2 in the light. In the morning, the products of the reductive pentose phosphate cycle gained the bulk of radioactivity, and malate was only slightly labeled. At the end of the light period, the opposite was true. From these results, it was deduced that during the early light period, the C_4 pathway of CO_2 fixation is blocked. At the end of the light period, it is active again, competing successfully with the reductive pentose phosphate cycle for exogenous CO_2 (see Chap. 4.2.3.2). However, it might be true that the changing label pattern observed may occur only under the relatively low light intensities (about 7000 lux) applied in Kluge's experiments.

It is well known that only in the light are significant amounts of carbohydrates synthesized (Table 3.2). The amount of malic acid accumulation in succulent tissue governs the extent to which sucrose or malate are produced. If endogenous malic acid is high, sugars are largely synthesized, and if malic acid is low, malic acid is largely synthesized (Kluge, 1971 b). The rate of exogenous CO_2 fixation in the light (but not in the dark) is apparently independent of the malic acid level (Kluge, 1971 b), however, there are contradictory data (Kluge, 1968 b).

With isolated mesophyll cells of *Sedum telephium*, Rouhani et al. (1973) were able to demonstrate ribulose diphosphate-dependent CO_2 fixation. Their further demonstration of P-enolpyruvate-dependent CO_2 fixation led them to conclude also that light CO_2 fixation could be through the PEP carboxylase sequence, or directly via ribulosediphosphate carboxylase.

Osmond and Allaway (1974) performed 15-s pulse $^{14}CO_2$ fixation experiments followed by a "chase" of $^{12}CO_2$ for several minutes. With *Kalanchoe daigremontiana* they found that during the beginning of the light period at the time of significant exogenous CO_2 uptake, malate had the labeling pattern of a direct β-carboxylation via PEP carboxylase, similar to the C_4-plant, *Atriplex spongiosa* (Osmond, 1967). ^{14}C-label was predominantly in the C-4 carboxyl of malate. During the steady-state light CO_2 fixation toward the end of the light period, malate was labeled more similarly to the C_3-plant, *Atriplex patula*. Hence, they concluded that the initial carboxylation was via RudP carboxylase followed by carboxylation of PEP by PEP carboxylase formed from PGA. Here labeling of malate is similar to the scheme proposed by Bradbeer et al. (1958) (see Chap. 3.1.5).

The above interpretation is supported by the finding of Osmond and Bjorkman (1975) that both dark CO_2 fixation and the initial burst of light CO_2 fixation were not affected by different O_2 concentrations. In contrast, CO_2 fixation at the end of the day was high with low oxygen concentration and vice versa (see Fig. 5.12). The conclusion is that only at the end of the light period does RudP carboxylase/oxygenase contribute substantially to fixation of exogenous CO_2.

3.2.4 Photorespiration in CAM Plants

There have been few specific studies of photorespiration in CAM plants (see Osmond, 1976 a, 1978), perhaps largely because of the complexities of their gas exchange patterns. The presence of photorespiration is partly indicated by glycolate synthesis in the light, O_2 inhibition of CO_2 fixation, and momentary shifts in gas exchange following light/dark or dark/light transitions. The photorespiration scheme as outlined by Tolbert (1971) and others is shown in Figure 3.10. The salient features are two sites for O_2 consumption, a site for CO_2 evolution in the light, the production of glycolate, and the O_2/CO_2 competition at the carboxylation site of ribulosebiphosphate carboxylase.

As early as 1956, Moyse and Jolchine obtained evidence for light-dependent glycolate synthesis in *Bryophyllum*. Hence, the substrate for photorespiration is produced in the light by CAM plants.

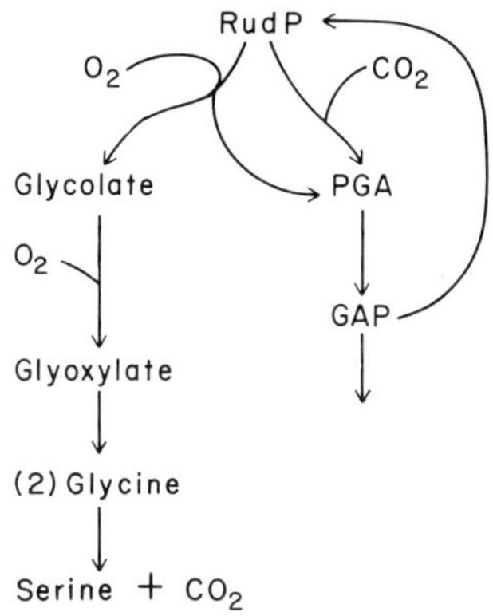

Fig. 3.10. Salient features of the proposed photorespiration pathway in green plants. Because of the oxygenase activity of RudP carboxylase, glycolate is produced at the expense of PGA. The glycolate is subsequently oxidized with concomitant oxygen consumption through glycolate oxidase. The glyoxylate thus produced is aminated to glycine; two of which can be converted to serine and CO_2. Hence, oxygen is consumed and CO_2 released in the light

Badger et al. (1975) reported that the RudP carboxylase isolated from *Kalanchoe daigremontiana* is not unlike that of C_3 plants (or C_4 plants) in that it shows oxygenase activity and is inhibited by O_2.

The first indication of a light/dark transition was in the work of Gregory et al. (1954) in their now classical study of gas exchange in *Kalanchoe blossfeldiana*. Subsequently, Despain et al. (1970) studied seedlings of *Carnegia gigantea* with an indication of photorespiratory gas exchange. Wynn et al. (1973) compared the postillumination CO_2 burst in CAM, C_4, and C_3 species. They found that CAM and C_4 plants were similar in that the amount of $^{14}CO_2$ released in 1 min following a light/dark transition seemed dependent on the recently fixed $^{14}CO_2$, whereas in the C_3 plants studied it was not. Although indicative of photorespiration, CAM responded differently than C_3 plants.

Kluge (1968a) reported that the $^{14}CO_2$ released in the light came from malate and hence, in part would be dissimilar to photorespiration. Szarek and Ting (1974a, b) also found that light CO_2 release increased with increasing acid level, also suggesting that the source of CO_2 evolution in the light was not entirely from the usual photorespiratory substrates. Similarly, O_2 evolution decreased with increasing acidity.

Osmond and Björkman (1975) found that light-dependent CO_2 fixation in *K. daigremontiana* was inhibited by O_2 in the same manner as C_3 plants, whereas dark CO_2 fixation was independent of 4%, 26%, or 36% O_2. Further, O_2 inhibition of photosynthetic CO_2 fixation is accompanied by a high CO_2 compensation point (Allaway et al., 1974a). Unlike C_3 plant photorespiration, however, the O_2 inhibition of photosynthetic CO_2 fixation is not eliminated by high CO_2 tension, and the postillumination CO_2 burst is present at low O_2 (Osmond and Björkman, 1975; see Fig. 5.13).

Crews et al. (1975) observed two postillumination CO_2 bursts in CAM plants. They have interpreted the initial burst as a consequence of photorespiration and a secondary burst as the result of malic acid decarboxylation (Crews et al., 1976). The hypothesis is based largely on the observation that the initial peak is reduced by low O_2 or high CO_2 whereas the second peak is not, and the second peak correlates with PEP carboxykinase and acid levels.

We can conclude that CAM plants do have photorespiration in that glycolate is produced in the light, RudP carboxylase has oxygenase activity, there is a postillumination CO_2 burst, and O_2 inhibits photosynthesis. Dissimilarities between these photorespiratory features and photorespiration in C_3 plants are most likely due to the gas exchange phenomena centered around malate metabolism.

In *K. daigremontiana* the occurrence of microbodies, where parts of the photorespiratory pathway are localized, has been demonstrated by Kapil et al. (1975).

3.3 Carbon Isotope Composition

In the free atmosphere, the ratio of $^{12}C/^{13}C/^{14}C$ in CO_2 is about 1.1% ^{13}C and 10^{-10}% ^{14}C. It is now well known that plants do not assimilate the isotopes of carbon equally, but discriminate against the heavier isotopes (Bender, 1968, 1971; Smith and Epstein, 1971). Carbon dioxide assimilation therefore results in carbon isotope fractionation. The isotope composition is usually expressed as

$$\delta^{13}C(^0/_{00}) = \left[\frac{^{13}C/^{12}C\ \text{sample}}{^{13}C/^{12}C\ \text{standard}} - 1\right] \times 10^3 .$$

Based on CO_2 from a carbonate standard (PDB), which is the fossil skeleton of one *Belemnitella* found in the Peedee formation of South Carolina, air over the oceans where contribution of CO_2 from plant and human activity is negligible has a $^{13}C/^{12}C$ relative ratio ($\delta^{13}C$) of −7‰. Referring to this PDB standard, C_3 photosynthetic plants have a $\delta^{13}C$ mode of −27‰ and C_4 photosynthetic plants a mode of −11‰ (Lerman, 1975). Hence, carbon assimilation by C_4 plants occurs with less discrimination or fractionation of ^{13}C (and also ^{14}C) than C_3 plants. Therefore, C_3 plants discriminate on the average 1.6‰ more than C_4 plants (see model by Lerman et al., 1974). It was shown by Whelan et al. (1973) that the carboxylating enzymes, P-enolpyruvate carboxylase and ribulosephosphate carboxylase, were responsible for a large part of the fractionation; P-enolpyruvate carboxylase from *Sorghum* discriminates less against ^{13}C than the other carboxylating enzymes. Similar experiments were conducted by Deleens et al. (1975) with both RudP carboxylase and PEP carboxylase isolated from spinach and maize, and although the data were quantitatively different from that of Whelan et al., PEP carboxylase discriminated less against ^{13}C than RudP carboxylase. Hence, the observation of Whelan et al. was substantiated. A range of recent research now has shown that CAM plants have a variable $\delta^{13}C$ value, despite the extreme constancy for C_3 and C_4 plants (Fig. 3.11). CAM plants have been reported to vary from −14 to −31‰ (see Lerman, 1975).

Osmond et al. (1973) were able to shift the $\delta^{13}C$ from C_4-like values to C_3-like values in *Kalanchoe* sp. by increasing the diurnal temperature fluctuation (*K. daigremontiana*) or shifting from long to short photoperiods (*K. blossfeldiana*). Nal-

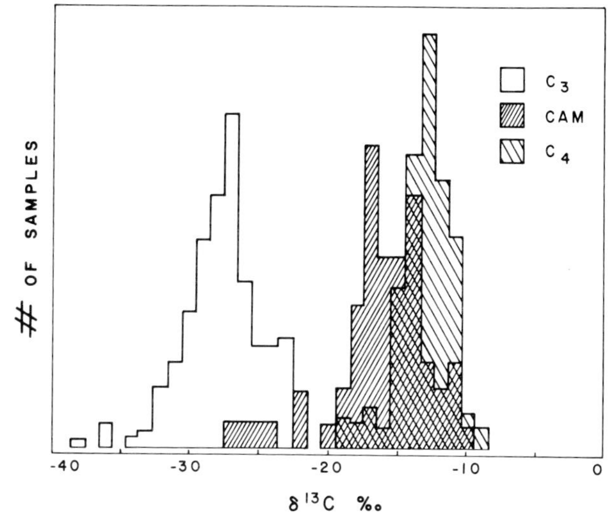

Fig. 3.11. Bar graph showing the ^{13}C composition of several C_3, C_4, and CAM species. Based on a fossil carbonate standard, the mode $\delta^{13}C$ of C_4 plants is $-11^0/_{00}$ and for C_3 plants it is $-27^0/_{00}$. Unpolluted air over oceans has a $\delta\,^{13}C$ of $-7^0/_{00}$. CAM plants, because of a variable contribution from PEP carboxylase and RudP, have a variable $\delta\,^{13}C$ estimate (redrawn from data of J.C.Lerman, 1975, by permission)

borczyk et al. (1975) and Osmond et al. (1976) found shifts of $\delta^{13}C$ in *K. daigremontiana* when water-stressed. Under continuous light, the $\delta^{13}C$ estimate was −25.9‰, suggesting CO_2 fixation through the C_3 pathway.

From the studies available in the literature, it has become apparent that the ^{13}C composition of CAM plants is highly variable, and depends highly on the environment. It is thus possible to make some predictions about the environmental conditions during growth of a CAM plant knowing the ^{13}C isotope composition (see Chap. 6.2.2).

Deleens et al. (1974) have proposed a tentative model to explain the variable isotope composition of CAM plants when grown in an atmosphere of $\delta^{13}C = -7‰$. In the case of C_3 plants, mean isotopic composition estimates of −27‰ result mainly because of the RudP carboxylase discrimination. C_4 plants with a coupled carboxylation reaction discriminate 4‰ on the average at the PEP carboxylase level, since all of the $^{13}CO_2$ is used by RudP carboxylase, there can be no additional discrimination and the final mean value of the carbon isotope composition is −11‰. With CAM plants, however, if they are assimilating CO_2 as C_3 plants through the RudP carboxylase pathway, then a ^{13}C discrimination of −20‰ is predicted, thus an isotope composition of −27‰. When assimilating first through the PEP carboxylase pathway, the isotope composition of the malate is −11‰, but the final $\delta^{13}C$ value in photosynthetic products will vary from −11‰ to −31‰ depending on whether all the CO_2 is refixed (−11‰, the same as for C_4 plants) or if there is complete refractionation by RudP carboxylase (−31‰). The final minimum value is −31‰ rather than −27‰, since fractionation starts from carbon with a $\delta^{13}C$ of −11‰ rather than the atmosphere −7‰

value. Mixing or losses of CO_2 from malic acid or the atmosphere will result in values between $-11‰$ and $-31‰$.

A very eloquent analysis of ^{13}C composition in CAM metabolism isolated from *Kalanchoe daigremontiana* leaves has recently been provided by Deleens and Garnier-Dardart (1978). These authors interpreted their results in terms of two independent sources for the substrate of β-carboxylation: PEP arises from glycolytic breakdown of insoluble storage carbohydrates rich in ^{13}C, and endogenous CO_2 comes from the complete respiratory breakdown of soluble sugars rich in ^{12}C. Hence, two independent glycolytic pathways were postulated. The above hypothesis is summarized in the following scheme (from Deleens and Garnier-Dardart, 1978) allowing the calculation of $\delta^{13}C$ of malate in *K. daigremontiana*:

$$\boxed{3\,C} \quad + \quad \boxed{1\,C} \longrightarrow \boxed{4\,C}$$

$$\text{PEP} \quad + \quad (CO_2\ \text{resp.} + CO_2\ \text{atm.}) \rightarrow \text{malate}$$

Products of carbohydrate breakdown

$$3/4\ \delta^{13}C_{(PEP)} + 1/4\ \delta^{13}C_{(CO_2\,resp. + CO_2\,atm.)} = \delta^{13}C_{(malate)}$$

3.4 The Proposed Total Carbon Flow in CAM

Based on the arguments and evidence outlined above, the following general scheme of carbon flow for CAM is proposed.

As outlined earlier in this chapter, the initial carboxylation product of CAM is oxalacetate. The oxalacetate is immediately reduced to malate or aminated to aspartate. It is malate or, more properly in CAM, malic acid which tends to accumulate. More than likely, through decarboxylation of malate, pyruvate is formed, some of which is aminated to alanine. Hence, the initial stable products of CAM are firstly malic acid, and then aspartate and alanine (Fig. 3.12).

Other organic and amino acids such as citrate, isocitrate, α-ketoglutarate, succinate, glutamate, glycine, asparagine, and glutamine (see Table 3.1) are presumably synthesized through the usual intermediary metabolism reactions commonly known.

By and large, no carbohydrates are produced in the dark, these are mostly produced in plants via photosynthetic dependent reactions.

The main flow of carbon from glucan to malic acid and back to glucan is depicted in Figure 3.13.

In the dark, it is assumed that stored starch or other glucan is hydrolyzed to sugar-phosphates which are metabolized to PEP via glycolysis and perhaps to some extent by pentose metabolism. Carboxylation of PEP results largely from atmospheric CO_2 to give oxalacetate, which is immediately reduced to malate. Some oxalacetate will ultimately go into aspartate synthesis. Some of the malate will be decarboxylated to pyruvate (or perhaps PEP), where the 3-carbon frag-

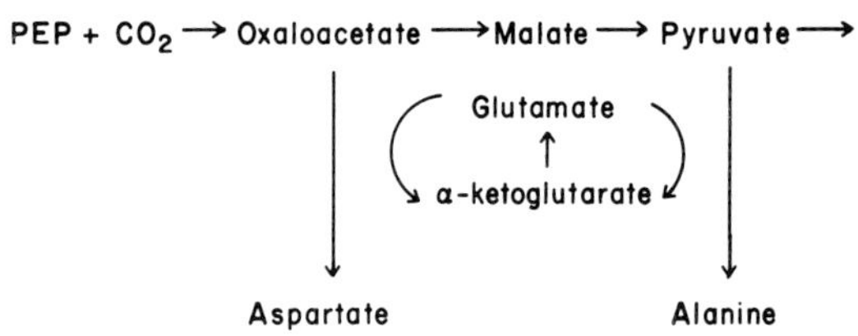

Fig. 3.12. Main reactions of CAM resulting in the primary product, malic acid, and secondary products aspartate, alanine, and pyruvate. Tertiary and other products are formed by the usual reactions of intermediary metabolism

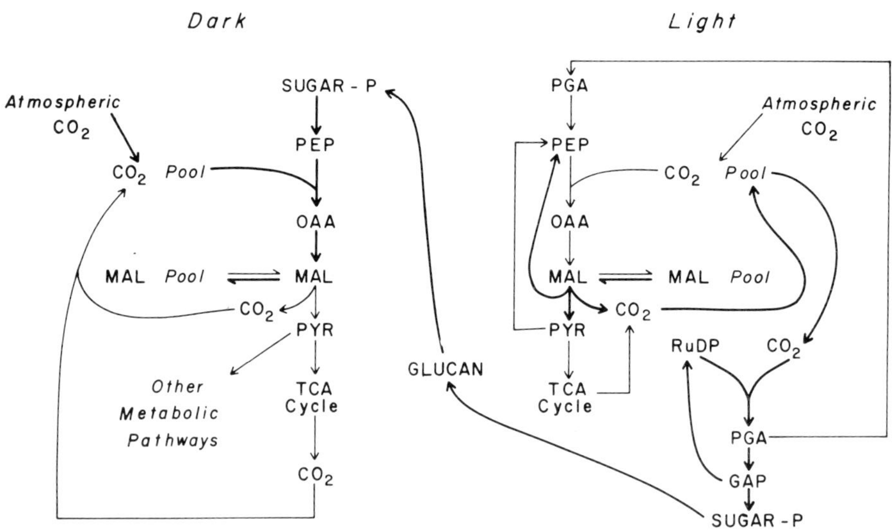

Fig. 3.13. The proposed metabolic pathway from storage carbohydrate to malic acid in the dark, and then to storage carbohydrate once again during the subsequent light period. Malic acid acts as a night storage molecule for CO_2 which is donated to the reductive pentose phosphate cycle during the subsequent day. The proposed gluconeogenesis starting from PEP in the light is not given in this scheme

ment may be metabolized in the citric acid cycle. Undoubtedly some of the cytoplasmic malate is randomized with citrate, succinate, and isocitrate in the citric acid cycle. The 3-carbon fragment could participate in alternate metabolic pathways such as lipid synthesis, but this is very low. The released CO_2 from malate decarboxylation and citric acid cycle activity may be recycled through the PEP carboxylase reaction, since respiratory gradients are extremely low.

In the light, events are largely reversed. The malate is decarboxylated to pyruvate or PEP, and the CO_2 enters the endogenous CO_2 pool. This CO_2 is then used as the substrate for ribulosediphosphate carboxylase to form phosphorylated sugars and ultimately storage carbohydrates. Hence, the storage carbohydrate, largely glucan, increases at the expense of malic acid. To some extent, atmospheric CO_2 can enter the CO_2 pool and ultimately be used as the substrate for sugar synthesis. If malic acid is high and contributing CO_2 to the pool, then atmospheric CO_2 goes into sugar. However, when malic acid is low, atmospheric CO_2 goes into malic acid.

The 3-carbon fragment, pyruvate or PEP, largely is converted back to PGA and then to stored carbohydrate. Some may, however, be metabolized by the citric acid cycle directly.

The above outline appears to be qualitatively correct, based on the available information in the literature. Much information needs to be gained on the quantitative flow of carbon in routes other than the

$$\text{glucan} \rightarrow \text{PEP} \rightarrow \text{malate} \rightarrow \text{glucan} \tag{3.10}$$

pathway. Also, it seems reasonable that since so many different and diverse plant families have CAM species, that there will be both quantitative and qualitative differences.

3.5 Comparison of CAM with Other Carboxylation Pathways in Plants

3.5.1 The Nonautotrophic C_4 Pathway of CO_2 Fixation

The nonautotrophic CO_2 fixation pathway (Fig. 3.14, see also section *Terminology*) functions largely in the synthesis of malate for ionic balance and osmotic adjustment (see e.g., Bernstein, 1963; Raschke, 1975a; Smith and Raven, 1976; Osmond, 1976b).

Further the malate may be decarboxylated to pyruvate with the subsequent generation of NADPH (Ting and Dugger, 1966; Peak et al., 1973). This pathway, therefore, results in an effective transhydrogenation of the H from NADH to NADPH at the expense of one ATP equivalent. In the direct pathway from P-enolpyruvate kinase.

$$\text{PEP} + \text{ADP} \rightarrow \text{pyruvate} + (\text{ATP}) \tag{3.11}$$

the ATP would be conserved. Here it is assumed that the PEP comes from the glycolytic sequence and the pyruvate, after conversion to acetate, can be used as a substrate for the citric cycle or in fatty acid synthesis. We visualize that in the case of fatty acid synthesis the NADPH from malate decarboxylation is used directly (Peak et al., 1973).

In comparison with CAM, the pathway to malate appears to be quite similar with perhaps the exception of the insoluble carbohydrate generator (Sutton, 1974a).

In CAM, the switch from malate production to malate consumption is largely governed by the light-dark cycle, whereas in the nonautotrophic pathway the flow through to pyruvate would be governed by the extent to which malate is used for ionic and osmotic adjustment, and NADPH is required for biosynthetic or other reactions.

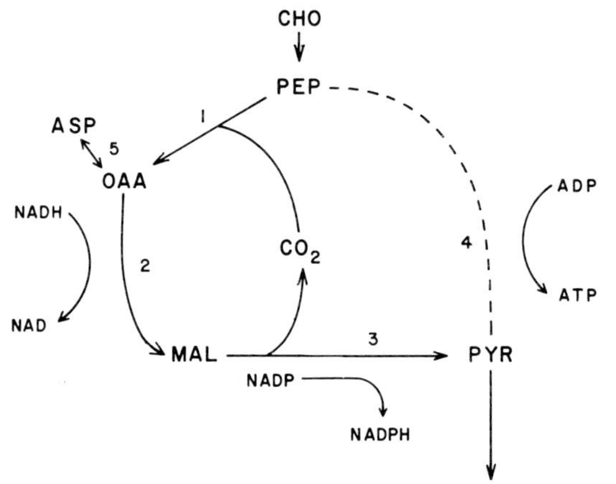

Fig. 3.14. The nonautotrophic CO_2 fixation pathway from PEP (probably generated through glycolysis) to malate and then to pyruvate. This essentially represents a bypass around the pyruvate kinase reaction producing an NADPH from an NADH at the expense of an ATP

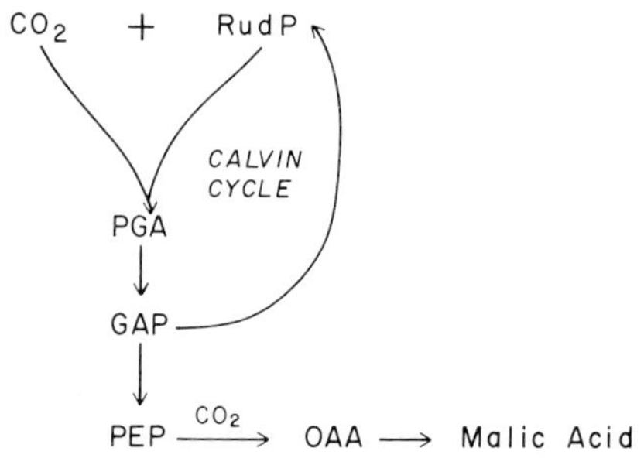

Fig. 3.15. The proposed production of malate during C_3 photosynthesis or CAM plant photosynthesis in the light. The PEP for carboxylation comes from photosynthetically produced PGA resulting in malate with a 2 to 1 ^{14}C labeling pattern of C-4 to C-1 (see Fig. 3.4)

A major difference between the dark nonautotrophic carboxylation sequence and CAM is that in the latter it is malic acid which is ultimately stored, whereas in the former, malate appears to accumulate.

3.5.2 C_3-Photosynthesis

Early photosynthetic research by Melvin Calvin's group at Berkely and Hans Gaffron in Florida clearly showed that malate was an important compound (Calvin et al., 1951; Gaffron and Fager, 1951). A question in the early 1950s was whether or not malate was an intermediate in the photosynthetic cycle, or if it was an end product? Extensive studies using algae and other C_3-photosynthetic plants clearly showed that malate was an end product. Malate was ruled out as an intermediate because unlike phosphoglycerate it did not come to rapid steady state and malonate-inhibited malate production, but did not interfere with total

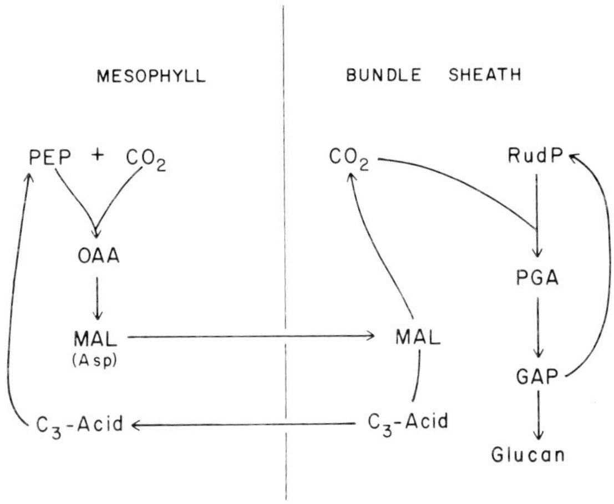

Fig. 3.16. The presumed C_4 photosynthetic pathway showing some of the similarities and differences between C_4 and CAM. Whereas in CAM malate is produced largely at night, in C_4 photosynthesis malate synthesis is light-dependent and proceeds in cells different from malate or C_4-acid decarboxylation. In C_4, carboxylation and decarboxylation are separated spatially whereas in CAM they are separated in time. Also, in C_4, PEP is generated from the C_3 product of the C_4-acid decarboxylation whereas in CAM it comes from stored carbohydrate synthesized during a previous light period

CO_2 fixation. Since the first compounds labeled in a $^{14}CO_2$ incubation experiment were phosphoglycerate, P-enolpyruvate, and C-4 acids, it was assumed that the C-4 acids came from carboxylation of P-enolpyruvate (see also Böcher and Kluge, 1977, 1978). The only apparent similarity between C_3 photosynthesis and CAM, therefore, is that the production of malate from PEP utilizes the same enzymatic mechanism (Fig. 3.15). In CAM, malate is clearly an intermediate.

When certain CAM plants feature the C_3-type of photosynthesis (i.e., when they are not doing CAM!, see e.g., Winter and Lüttge, 1976; Kluge, 1977b), by definition we would assume the photosynthetic pathway to be identical to C_3-photosynthesis.

3.5.3 C_4-Photosynthesis

Shortly after the elucidation of the C_4-photosynthetic sequence,

$$CO_2 + PEP \rightarrow MAL \rightarrow RudP \rightarrow PGA \qquad (3.12)$$

there was a marked renewed interest in CAM metabolism.

The similarities between CAM and C_4-photosynthesis are so great that W. M. Laetsch was prompted to state at the 1971 Photosynthesis and Photorespiration Conference in Canberra that "C_4-photosynthesis is CAM with Kranz." C. C. Black in his 1973 *Annual Review of Plant Physiology* article stated that CAM was a photosynthetic option comparable to the C_3 and C_4 options. In 1971 and in 1972 we compared the similarities and differences between C_4-photosynthesis and CAM (Ting, 1971; Ting et al., 1972). The most apparent metabolic difference is

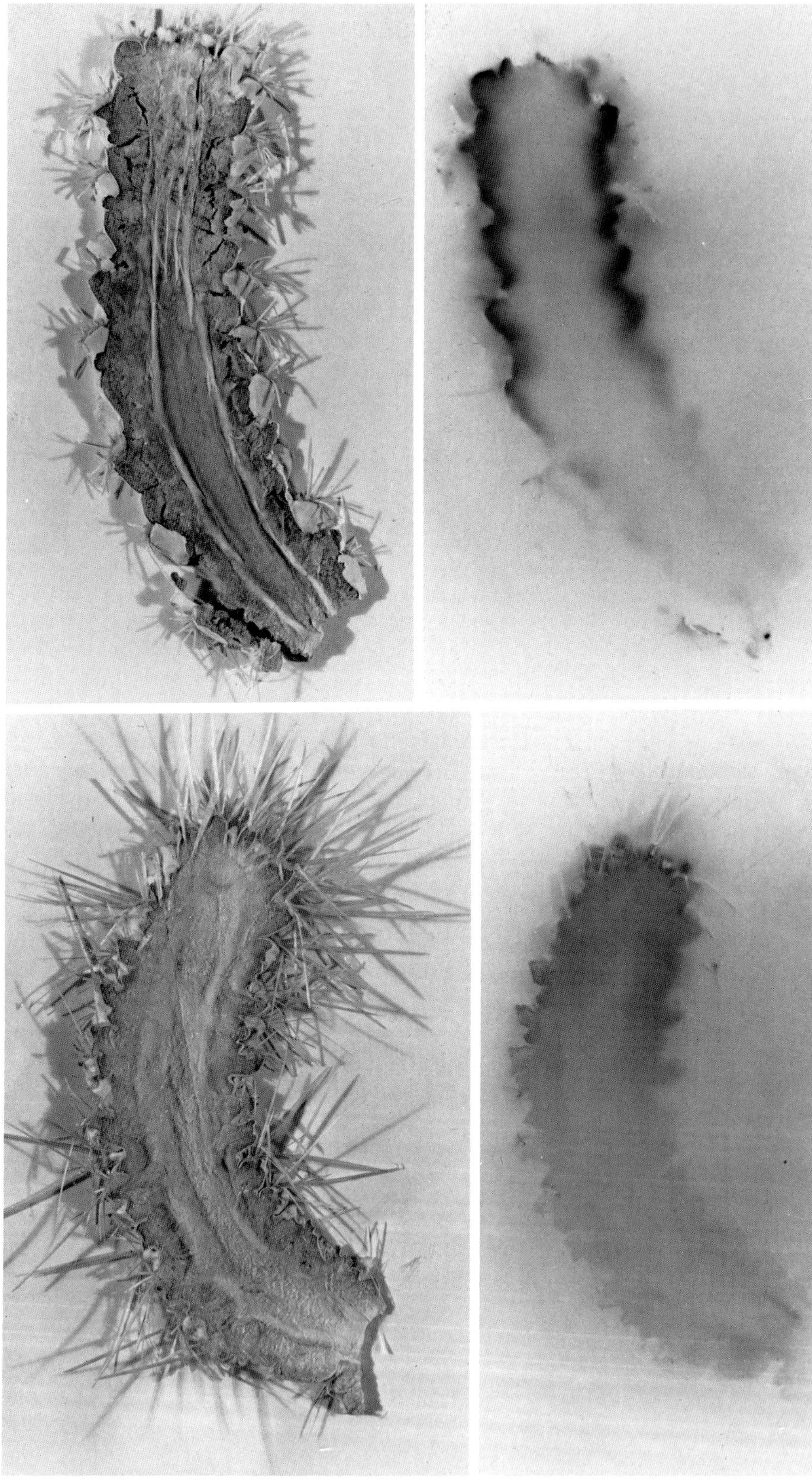

that in CAM the first carboxylation step to form malate and the second carboxylation step to form phosphoglycerate are separated in time (i.e., the night-day cycle) whereas in C_4 photosynthesis they are separated in space (i.e., mesophyll cells for the former and bundle sheath cells for the latter (Fig. 3.16). Other differences are subtle and apparently of a much lesser magnitude.

A functional difference between CAM and C_4-photosynthesis is the immediate source of P-enolpyruvate for carboxylation. In C_4-photosynthesis, it is largely assumed that PEP is regenerated from the 3-C fragment after malate (or oxalacetate) utilization. Two possibilities seem apparent: (1) pyruvate is converted to P-enolpyruvate by pyruvate, Pi dikinase (Hatch and Slack, 1970) or (2) oxalacetate rather than malate is decarboxylated to form P-enolpyruvate by PEP carboxykinase (Black et al., 1973).

3.5.4 Conclusions

These comparisons can be summarized by stating that CAM is functionally similar to C_4 plant photosynthesis except that in CAM the sequential carboxylations are separated temporarily, whereas in C_4-photosynthesis they are separated spatially. In favorable environments, certain CAM plants can exercise the C_3-photosynthetic option to exploit the environment.

Although the enzymatic machinery for P-enolpyruvate carboxylation in the nonautotrophic pathway, C_3-photosynthesis, C_4-photosynthesis, and CAM is similar, the enzymes are sufficiently different to be classed as isozymes, and hence, we assume evolutionary pressures have resulted in a common theme (PEP carboxylation) being modified for similar, but functionally different adaptive roles (see Chap. 4).

It should be noted that insofar as is known, the basic mode of net CO_2 fixation by all higher green plants is the same (Kelly et al., 1976). CO_2 fixation to produce net sugar in the initial form of glyceraldehyde phosphate takes place through the reductive pentose cycle outlined by Calvin and his co-workers. Alternate schemes such as CAM (and C_4-photosynthesis) could be more properly viewed as auxillary reactions (Kluge, 1971a, c; Kelly et al., 1976) giving environmental flexibility (see Chap. 6).

3.6 Translocation of CAM Products in the Plant

Little is known about translocation in CAM plants. In one study, Chow et al. (1966) exposed *Opuntia polyacantha* stem sections to $^{14}CO_2$ in the light for 30 min and assessed the amount of methanol-soluble ^{14}C-compounds in the exposed

◄
Fig. 3.17. Photographs of freeze-dried sections (*left*) of *Opuntia bigelovii* and autoradiographs (*right*) of the same sections after dark fixation of $^{14}CO_2$. *Upper:* 12 h after dark fixation; *lower:* 6 weeks after dark fixation. Plants were growing in situ at the P.L.Boyd Deep Canyon Desert Research Center, Palm Desert California. Little translocation out of the terminal stem sections occurred during the 6-week period, but radioactivity did migrate from the outer green tissue to the inner nongreen tissue (data obtained by B. Sutton and I. P. Ting, previously unpublished)

areas relative to distal areas. After 30 min approximately 5% of the total ^{14}C in the tissues was translocated. Soybeans were the same despite the fact that over 10 times more $^{14}CO_2$ was fixed. After 6 h, 4% of the total was in distal regions, as compared to 43% for soybeans. The *Opuntia*, however, had lost 77% of the total ^{14}C fixed after 0.5 h. Soybeans lost 52%. To what extent this loss represents conversion to insoluble components is not known. These data indicated little translocation of photosynthate in cactus.

A similar study of dark $^{14}CO_2$ by Sutton with *Opuntia bigelovii* yielded the same conclusion. Initially, virtually all the ^{14}C was in the stem section that was exposed, and further the activity was just present in the outer chlorophyll bearing tissue. After 6 weeks, little translocation had occurred (Fig. 3.17). These data indicate that in cactus at least, translocation of CO_2 fixation products is less than in nonsucculent plants, and could account in part for slower growth rates.

As indicated earlier (Chap. 2.3), Kluge and Knapp (unpublished) showed that carbon fixed in the CAM-performing outer green tissue of *Aloe arborescens* was retained in the green leaf tissue and being transported very slowly into the internal water-storing tissue. In further studies, Kluge and Konrad (unpublished) labeled a single phyllodium of *Kalanchoe tubiflora* attached to the plant by allowing $^{14}CO_2$ dark fixation. If the plant remained in the dark, radiocarbon was not exported from the labeled phyllodium. However, substantial label was transported first to the top of the plants and later to other parts if the plant was illuminated for some hours. This showed that malic acid synthesized in CAM is immobile, and carbon bound previously in malic acid is transported in the phloem only after conversion in photosynthesis to carbohydrates. In fact, the bulk of ^{14}C exported to the top of the plant was found in soluble and insoluble carbohydrates.

4. Control and Modification of CAM

4.1 Definitions

Perhaps the most significant feature of CAM is the regular, rhythmic diurnal cycle in which biochemical and physiological events follow very closely in time. Furthermore, some of the CAM activities may undergo alterations correlated with the seasonal time of the year. This chapter considers how the regularity of the diurnal and seasonal cycles are controlled, and how the basic CAM cycle is modified by external factors and pertubations from unusual light/dark periods, unusual concentrations of CO_2, and inhibitors. Recent reviews dealing in particular with CAM control have been published by Queiroz (1974) and Kluge (1976b, 1977a).

4.2 Metabolic Control of CAM

4.2.1 The CAM Enzymes

There is increasing evidence that CAM is controlled to a great extent at the enzyme level. As outlined earlier (see Chap. 3), the main metabolic sequence linking malate to the reductive pentose cycle is:

$$CO_2 + PEP \rightarrow OAA \rightarrow MAL\,(ASP) \rightarrow PYR\,(PEP) + CO_2 . \qquad (4.1)$$

Hence, those enzymes involving PEP carboxylation and regeneration, and malate production and degradation plus synthesis of the amino acids aspartate and alanine from the keto acids oxalacetate and pyruvate are deemed most interesting for our understanding in CAM.

4.2.1.1 P-Enolpyruvate Carboxylase [Orthophosphate: Oxalacetate Carboxylase (Phosphorylating)]

a) General Properties

P-enolpyruvate carboxylase was first observed and measured in spinach leaf extracts by Bandurski and Greiner (1953), and was quickly shown to be an important carboxylating enzyme in succulent plants (e.g., Walker, 1957). Since its discovery and implication in Crassulacean Acid Metabolism, it has been studied

Table 4.1. Activities of several CAM enzymes from *Kalanchoe daigremontiana*[a]

Enzyme	μmol min^{-1} mg chl^{-1}
PEP carboxylase	19.2
Malate enzyme	1.3
Aspartate amino transferase	15.1
Glycine pyruvate transaminase	2.3
Malate dehydrogenase (NAD)	183.0
Malate dehydrogenase (NADP)	1.8

[a] Data obtained at Australian National University, Canberra, 1972.

in a variety of succulent plants including *Kalanchoe* (Table 4.1), *Bryophyllum*, *Sedum*, *Agave*, *Sempervivum*, *Opuntia*, and *Nopalea* (see, e.g., papers by Garnier-Dardart, 1965; Brandon, 1967; Mukerji, 1968; Queiroz, 1968a, b; Mukerji and Ting, 1968b; Kluge and Osmond, 1972; and Ting and Osmond, 1973a–c). The protein has not been purified from succulent plants, but was prepared to near homogeneity from potato tuber tissue and peanut cotyledons by Smith (1968) and Maruyama et al. (1966).

The enzyme catalyzes the reaction

$$PEP + HCO_3 \rightarrow OAA + iP. \tag{4.2}$$

It can use Mg^{2+} or Mn^{2+} as activators. The plant enzyme can readily be separated from other metabolic proteins by techniques of ionic salting and ion exchange or electrophoretic chromatography (Ting and Osmond, 1973a, b).

Molecular weights of PEP-C from nonsucculent plants are reported to be about 350,000 (Maruyama et al., 1966; Smith, 1968; Ting and Osmond, 1973a, b). Enama (1976) compared four species of C_4 plants and found a value of 340,000 for

Table 4.2. Molecular weights of PEP carboxylase from species of the genus *Sedum*[a]

Species	Mol. weight of PEP-C	$\delta\,^{13}C$ (‰)	Mode of photosynthesis
Sedum acre	136,000	−28.06	Facultative CAM
Sedum coeruleum	162,000	−32.66	C_3 [b]
Sedum compressum	162,000	−18.66	CAM
Sedum morganianum	172,000	−14.61	CAM
Sedum praealtum	136,000		CAM [c]
Sedum sieboldii	fraction a 136,000 fraction b 162,000		C_3 [b]
Sedum spectabile	162,000	−25.5	CAM [c]
Sedum telephium	172,000	−28.0	CAM [c]

[a] Data obtained by Knopf and Müller in Kluge's laboratory.
[b] The species showed C_3 mode of photosynthesis when investigated. Facultative CAM cannot be excluded.
[c] The plants showed dark CO_2 fixation and acid oscillation. Substantial amounts of CO_2 were also taken up during the day.

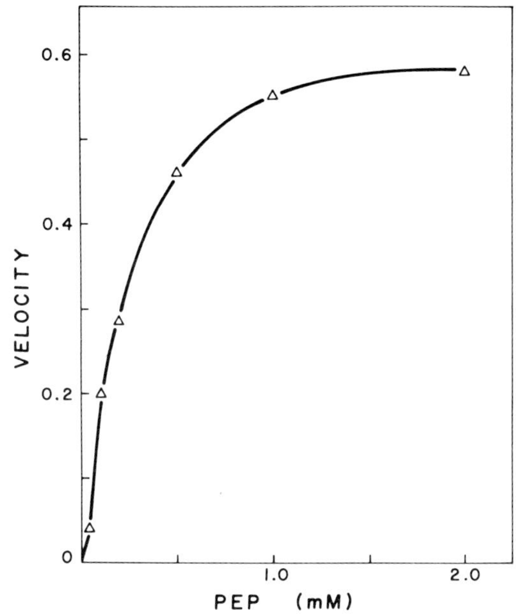

Fig. 4.1. Typical kinetic rate curve for *Kalanchoe tubiflora* PEP carboxylase over substrate range of 0.02–2.0 mM PEP (previously unpublished)

Tidestromia oblongifolia PEP-C and about 180,000 for the PEP-C isolated from *Distichlis spicata*, *Atriplex sabulosa*, and *A. hymenelytra*. Knopf in Kluge's laboratory investigated several species of *Sedum* (see Table 4.2). All species with the exception of *S. sieboldii* showed a single electrophoretic protein band having PEP-C activity when electrophoresed on acrylamide gels. The molecular weights ranged from 136,000 to 172,000. *Sedum sieboldii* had two bands with estimated molecular weights of 136,000 and 162,000. The band with the estimated size of 172,000 isolated from *S. morganianum* dissociates into a band with molecular weight of 64,000 when re-run on SDS polyacrylamide gels (Müller in Kluge's laboratory, unpublished). In the green tissue of *Aloe arborescens*, a PEP-C with molecular weight of about 340,000 was observed (Licht in Kluge's laboratory, unpublished).

It is of course not possible to say whether the 340,000 form is the in vivo form of PEP-C with subunits of about 170,000, or if the latter dimerizes to form 340,000 proteins. It would seem from these data that the minimum subunit size is about 64,000.

A typical P-enolpyruvate-dependent rate curve for *Kalanchoe tubiflora* is shown in Figure 4.1. The curve is quite hyperbolic with little or no tendency toward sigmoidicity. Despite this, there is evidence for allosteric properties of plant PEP carboxylase (Coombs et al., 1973; Ting and Osmond, 1973a; Mukerji, 1974).

Estimates of maximum velocity of the enzyme per minute per mg chlorophyll range from 4 µmol oxalacetate produced min^{-1} mg $chlorophyll^{-1}$ to over 70 µmol min^{-1} mg $chlorophyll^{-1}$. The mean for seven species was about 35

(Kluge and Osmond, 1972; Ting and Osmond, 1973a, b). Because of endogenous inhibitors, diurnal and seasonal fluctuations of enzymic activity, and environment-dependent shifts in metabolic pathways, maximum enzymatic velocities in succulent plants are difficult to obtain and awkward to evaluate. Nevertheless, maximum velocity estimates are high and in the C_4 plant range of 30 μmol min^{-1}, and far above the estimate of 1.5 μmol min^{-1} mg $chlorophyll^{-1}$ of C_3 plants.

Estimates of apparent Km for P-enolpyruvate are low with a mean of about 0.2 mM at pH 7.8 (Ting and Osmond, 1973a, b). As is now well known, the Km varies significantly with ionic strength and pH, but the data are of value, since they indicate that the enzyme is functional at low substrate levels. Reports in the literature indicate that the Michaelis constant range for magnesium is 0.1–0.5 mM in the same range as for PEP (Ting and Osmond, 1973a, b). In this regard, it is not clear what the actual substrate for the reaction is, i.e., free PEP or a PEP-metal ion complex. Estimates of the Km for total CO_2 are about 0.02 mM (Ting and Osmond, 1973a, b). The "CO_2" substrate is most probably bicarbonate rather than CO_2 (see Chap. 3.1.4). The phosphate acceptor is water.

P-enolpyruvate carboxylase is an extremely efficient carboxylating protein because of the large change in free energy accompanying P-enolpyruvate hydrolysis on the order of −12 Kcal per mol.

The thermodynamic equilibrium constant for the malate dehydrogenase reaction is about $1.7 \cdot 10^{12}$ (Yoshida, 1965) and hence, the overall reaction [see Eq. (4.3)] is practically not reversible. It is not surprising, therefore, that succulent tissues with these enzymes tend to accumulate massive quantities of malic acid:

$$PEP + CO_2 \rightarrow \text{Malate}. \quad (4.3)$$

b) Regulatory Properties

Because of the interest in the control of CAM metabolism, the regulatory properties of P-enolpyruvate carboxylase as the key enzyme are important.

In microbial systems, P-enolpyruvate carboxylase is reported as an allosteric enzyme activated by acetyl-CoA and severely inhibited by aspartate and malate (Canovas and Kornberg, 1966; Maeba and Sanwal, 1965).

Table 4.3. Effect of L-malate and glucose-6-P on PEP carboxylase of *Kalanchoe daigremontiana* kinetics

	V_{max} [a]	R [b]	Km [c]	R [b]
Control	0.42	1.0	0.13	1.0
+1 mM G-6-P	0.45	1.07	0.08	0.61
+2 mM G-6-P	0.47	1.12	0.05	0.39
+5 mM l-malate	0.31	0.74	0.21	1.6
+5 mM l-malate and 2 mM G-6-P	0.34	0.81	0.05	0.39

[a] V_{max} = Maximum velocity determined by double reciprocal plots.
[b] Ratio of experimental to control.
[c] Km = Michaelis constant for PEP, mM.

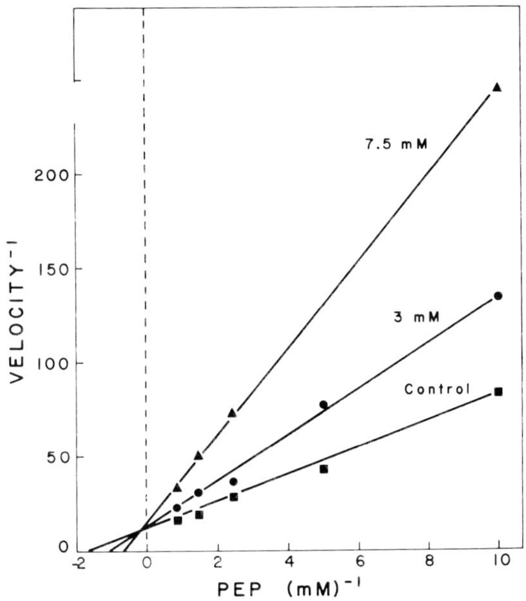

Fig. 4.2. Double reciprocal graphical analysis of the malate inhibition of *Byrophyllum* PEP carboxylase. The reciprocal of the initial velocity is plotted against the reciprocal of PEP concentration of 3, and 7.5 mM malate. The estimated K_i for malate is 3.6 mM at pH 6.25 (data of Kluge and Osmond, 1972)

Malate Inhibition (see Table 4.3). Following our initial observation that L-malate and oxalacetate, and to a lesser extent D-malate, aspartate, and glutamate inhibit maize root PEP carboxylase (see Ting, 1971), Queiroz (1967) reported that the PEP carboxylase of *Kalanchoe* leaves was sensitive to L-malate. Based on these data, it was proposed that L-malate could regulate its own synthesis by feedback inhibition of PEP carboxylase (Queiroz, 1967). It is possible that malate is simply a product analog, acting similarly to oxalacetate, yet in maize roots, oxalacetate and malate inhibit in a different manner (Ting, 1968a).

Kluge and Osmond (1972) reported malate and pyruvate inhibition of partially purified PEP carboxylase from *Kalanchoe tubiflora* and *Sedum praealtum* with a Ki for malate of 3.6 mM at pH 6.25 (Fig. 4.2). Since malate may accumulate to a level of 100 μmol per g fresh weight in crassulacean plants, 1% of this concentration in the vicinity of PEP carboxylase would be sufficient for inhibition.

Glucose-6-Phosphate Activation. Recently, it was reported that low concentrations of glucose-6-phosphate, but not other phosphorylated sugars, could activate PEP carboxylase (Ting and Osmond, 1973c; Coombs et al., 1973). The data suggest, because of the nature of activation, that plant PEP carboxylase could be an allosteric protein.

We reported that glucose-6-phosphate reduced the apparent Km-PEP of *Kalanchoe daigremontiana* PEP carboxylase but did not increase the theoretical maximum velocity (Ting and Osmond, 1973c). As shown in Figure 4.3, glucose-6-phosphate reduces malate inhibition, and hence we visualize regulation of PEP

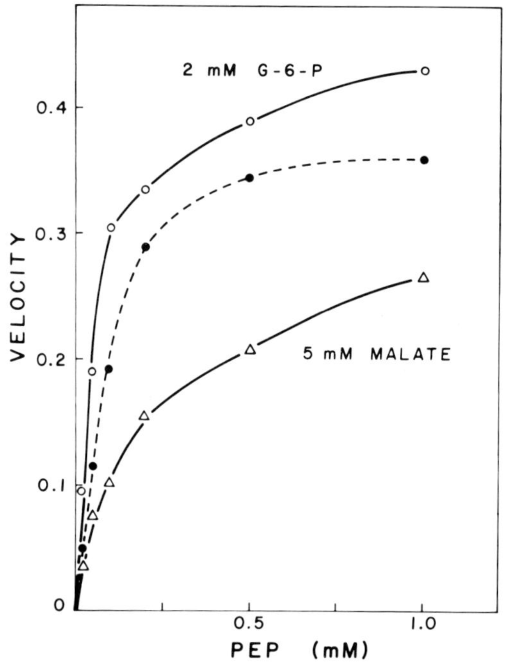

Fig. 4.3. Interaction of 2 mM glucose-6-phosphate (G-6-P) and 5 mM malate in modulation of *Bryophyllum* PEP carboxylase. G-6-P increases the activity and malate decreases. In combination, activity is largely restored to the control ●---●, assay without inhibitor or activator (Ting and Osmond, 1973c)

carboxylase malate synthesis by glucose-6-phosphate/malate levels in CAM plants.

Other Metabolites as Effectors. A variety of other naturally occurring compounds inhibit the PEP carboxylase reaction, including organic and amino acids (Mukerji and Ting, 1971; Ting, 1968a; Lowe and Slack, 1971; Kluge and Osmond, 1972). In most cases, concentrations of inhibitors are high (5–30 mM) and the physiological significance is not clear.

Table 4.4. Inhibition of PEP carboxylase by chelating compounds[a]

Substance	Inhibition (%)
ATP	86
ADP	48
AMP	0
Citrate	57
EDTA	100

[a] PEP carboxylase purified from *Atriplex spongiosa*. Reaction run at saturating PEP concentration and Km for Mg. Chelators at 1 mM.

Table 4.5. Effect of NaCl on PEP carboxylases of *Atriplex hastata* and *Atriplex spongiosa*. V_{max} values were estimated from double reciprocal plots and expressed relative to 0 mM NaCl. The interaction coefficients n were determined from log-log graphs of the Hill equation [a]

	NaCl (mM)			
	0	10	50	100
Atriplex spongiosa				
V_{max}	1.00	0.98	0.98	0.93
Km	0.49	0.8	1.21	2.0
n	1.25	1.22	1.25	1.21
Atriplex hastata				
V_{max}	1.00	—	—	1.07
Km	0.08	—	—	0.275
n	0.78	—	—	0.76

[a] From Ting and Osmond, Plant Physiology *51*, 443 (1973).

Metal Chelators. Naturally occurring and synthetic substances which can chelate manganese or magnesium tend to be very strong inhibitors of PEP carboxylase. At Km levels of magnesium, 1 mM EDTA will inhibit nearly 100%, as will citrate. Nucleotides such as ATP and ADP, but not AMP will also inhibit at 1 mM concentrations (Table 4.4). One can readily visualize therefore a possible regulatory function for ATP and ADP in tissues with PEP carboxylase-mediated reactions.

Inorganic Anions. Inorganic anions, most notably chloride, severely inhibit PEP carboxylase (Osmond and Greenway, 1972). The inhibition is most severe when the reaction is run at the apparent PEP Km. Detailed studies show that the theoretical maximum velocity for the reaction is not significantly affected, but the apparent Km for PEP is increased as much as four fold in the presence of 100 mM NaCl. Hence, it appears that chloride interferes with the binding of PEP to the enzyme (Table 4.5).

Other inorganic anions including sulfate, phosphate (Ting and Osmond, 1973a, b), and sulfide inhibit, the latter presumably by competition with bicarbonate (Avadhani et al., 1971).

Von Willert (1975b, c) reported that inorganic phosphate added in concentrations of about 30 mM to in vitro assay systems reduced the sensitivity of *Mesembryanthemum crystallinum* PEP-C to malate and NaCl (see Chap. 4.3.3).

Mukerji and Ting (1971) have shown PEP carboxylase to be inhibited by high concentrations of CO_2 and bicarbonate.

4.2.1.2 Malate Dehydrogenase (L-malate: NAD Oxidoreductase)

There is a NADP-specific MDH known to occur in plants. It was characterized by Johnson and Hatch (1970) from maize and by Ting and Rocha (1971) from spinach. The latter has an estimated molecular size of 110,000, but breaks into smaller units in the absence of dithiothreitol. The apparent Km's for oxalacetate and L-malate were 0.04 and 11.0 mM respectively.

The reaction favors malate with an estimated equilibrium constant based on initial rates of $2 \cdot 10^3$.

The NADP-specific MDH is known to be present also in CAM plants (Table 4.1). The role of this enzyme in CAM remains to be evaluated.

NAD-malate dehydrogenase catalyzes the reversible reduction of oxalacetate

$$\text{OAA} + \text{NADH}_2 \rightarrow \text{NAD}^+ + \text{Mal}. \tag{4.4}$$

The NAD malate dehydrogenase-mediated reaction strongly favors reduction of oxalacetate. Michaelis constants for oxalacetate are 1–2 orders of magnitude greater than for L-malate. Relatively little is known about the regulation of plant NAD-malate dehydrogenase, but the MDH of the Enterobacteriaceae is regulated by NADH in an allosteric manner (Sanwal, 1969). In plant tissues, the freshly prepared malate dehydrogenase isoenzymes are inhibited by the substrate, oxalacetate. The three known isoenzymes, soluble, mitochondrial, and microbody, are affected differentially; however, the soluble and microbody proteins are more sensitive to pH changes being inhibited more by oxalacetate at low pH than at high pH. The mitochondrial form is inhibited about equally at all pH values (Yamazaki and Tolbert, 1969; Rocha and Ting, 1970; Ting, 1968b). When purified, the substrate inhibition properties are lost.

A variety of small, naturally occurring metabolic substrate and product molecules inhibit and otherwise affect the activity of *Opuntia* malate dehydrogenase (Mukerji and Ting, 1969). For example, α-ketoglutarate, *cis*-aconitate, isocitrate, succinate, fumarate, pyruvate, aspartate, glutamate, and asparagine tend to inhibit all isoenzymes. The affect is much greater in the direction of malate oxidation than oxalacetate reduction. Inhibitor concentrations had to be relatively high (10 mM or more) and no obvious differential effects were noted. The physiological significance of these effects is not clear at this time.

Because the NAD-malate dehydrogenases are such powerful catalysts, substrate level regulation is not likely. It is more reasonable to assume that it keeps the unstable, metabolically active keto acid (oxalacetate) in the stable form of malate.

The NADP-specific malate dehydrogenase is known to be light-activated (Hatch and Slack, 1969; Johnson and Hatch, 1970), and indeed must be prepared and assayed in the presence of dithiothreitol or other powerful sulfhydryl reagents. In spinach tissue, we observed that the naturally occurring lipoate would also activate and could be an in vivo regulator of this enzyme. Since the protein is apparently light-activated, its role in dark CO_2 fixation is nil.

4.2.1.3 Aspartate Aminotransferase (L-aspartate: α-oxoglutarate Aminotransferase)

Aspartate aminotransferase or glutamate-oxalacetate transaminase catalyzes the following readily reversible reaction:

$$\text{GLU} + \text{OAA} \rightarrow \alpha\text{OG} + \text{ASP}. \tag{4.5}$$

The reported molecular weight is about 110,000 and the enzyme has 2 mol pyridoxal phosphate per mol of enzyme. It is reasonable to assume that the aspartate formed during CAM metabolism is by glutamate transamination to oxalacetate.

4.2.1.4 "Malate Enzyme" [L-malate: NADP Oxidoreductase (Decarboxylating)]

"Malate enzyme" catalyzes the following reversible reaction:

$$MAL + NADP^{+} \xrightarrow{Mg^{2+}} NADPH_2 + PYR + CO_2 . \quad (4.6)$$

It is widely distributed in plant tissues and has been studied in succulent plants (Mukerji and Ting, 1968a; Brandon, 1967; Garnier-Dardart and Queiroz, 1974; Walker, 1960). It is assumed that in many succulents, malate decarboxylation is by malate enzyme.

Dittrich et al. (1973) surveyed 11 different plant families and reported substantial, but variable NADP-malate enzyme activity in all species studied.

The equilibrium constant (0.05 mol/l) favors malate formation. It is usually assumed that in vivo the reaction is toward malate utilization rather than synthesis (Walker, 1962). The activity is usually less than PEP carboxylase (Table 5.1).

Michaelis constants for L-malate are in the range of 0.1–0.7 mM (Dilley, 1966; Mukerjii and Ting, 1968a). Garnier-Dardart and Queiroz (1974) reported 0.82 mM for the malate Km (pH 6.9), and 0.006 mM for the NADPH Km with an enzyme prepared from *Kalanchoe daigremontiana*.

Recently, a malate enzyme using NAD rather than NADP as the redox co-factor has been reported in some CAM plants (Dittrich, 1975). Hence, there is the probability of malate decarboxylation in CAM plants using either NADP or NAD as a co-factor.

The regulation of malate enzyme is poorly understood. Although the equilibrium constant favors malate formation, the CO_2 release tends to drive the reaction toward decarboxylation (Harary et al., 1953). Hence the suggestion that the in vivo reaction is toward malate utilization rather than malate synthesis (Walker, 1962; Ting and Dugger, 1965).

In a discussion of the diurnal variation of organic acids in succulents, Queiroz (1968b) indicated that L-malate activated the malate enzyme. Kinetic studies of the apple fruit enzyme did not suggest substrate activation; however, D-malate enhanced the activity when in the presence of L-malate (Bandurski, 1955). The cactus isoenzymes were inhibited by D-malate (Mukerji and Ting, 1968a). Carefully controlled kinetic experiments designed to determine if there was any substrate (L-malate) activation were negative (Mukerji and Ting, 1968a). However, the enzyme from *Kalanchoe daigremontiana* showed positive cooperativity for malate (Garnier-Dardart and Queiroz, 1974).

4.2.1.5 P-enolpyruvate Carboxykinase [ATP: Oxalacetate Carboxylase (Transphosphorylating)]

P-enolpyruvate carboxykinase catalyzes the reversible decarboxylation of oxalacetate:

$$OAA + ATP \rightleftarrows PEP + CO_2 + ADP . \quad (4.7)$$

This has been variously demonstrated in plant tissues and surveyed in a variety of succulents by Dittrich et al. (1973) where its activity in some may be as high as 15 μmol min^{-1} mg chl^{-1}.

Of the 11 succulent families surveyed by Dittrich et al. (1973) the Asclepiadaceae, Bromeliaceae, Euphorbiaceae, Liliaceae, and Vitaceae had substantial PEP carboxykinase activity (a mean of about 8 μmol min^{-1} mg chl^{-1}). The remaining seven families (Agavaceae, Aizoaceae, Asteraceae, Cactaceae, Crassulaceae, Orchidaceae, and some members of the Liliaceae) showed little or no PEP carboxykinase activity.

Those plants with high PEP carboxykinase activity had low malate enzyme (1.2 μmol min^{-1} mg chl^{-1}) whereas those with no detectable PEP carboxykinase activity had higher levels of malate enzyme (4.2 μmol min^{-1} mg chl^{-1}).

Similarly to malate enzyme, despite the unfavorable free energy change, PEP is favored because of CO_2 release, and it is assumed that the enzyme functions in vivo in the direction of PEP formation.

In CAM metabolism, it is assumed that PEP carboxykinase may consume malate after conversion to oxalacetate by forming PEP directly.

4.2.1.6 Pyruvate, Phosphate Dikinase

Pyruvate, Pi dikinase is an enzyme encountered by Hatch and his co-workers during studies of C_4 photosynthesis (Hatch and Slack, 1969).

It catalyzes the reversible phosphorylation of pyruvate:

$$\text{ATP} + \text{PYR} + \text{Pi} \rightarrow \text{PEP} + \text{AMP} + \text{PPi}\,. \qquad (4.8)$$

It is known to be light-activated and in vitro requires a thiol reagent and Mg^{-2}. Its presumed role is the generation of PEP from pyruvate.

It was first reported to occur in CAM plants by Kluge and Osmond (1971), and Sugiyama and Laetsch (1975) studied it with immunochemical techniques from *Kalanchoe daigremontiana.*

4.2.1.7 Alanine Aminotransferase (L-alanine: α-oxoglutarate Aminotransferase)

Alanine aminotransferase catalyzes the reversible transamination of glutamate to pyruvate:

$$\text{GLU} + \text{PYR} \rightarrow \alpha\text{-OG} + \text{ALA}\,. \qquad (4.9)$$

It is known to occur in CAM plants and has been implicated in alanine production during CAM by Queiroz et al. (1972).

4.2.1.8 Ribulose-1.5-Bisphosphate Carboxylase/Oxygenase [3-phospho-D-glycerate Carboxylase (Dimerizing)]

$$\text{RudP} + CO_2 \rightarrow 2\,\text{PGA}\,. \qquad (4.10)$$

In a survey of some 35 different CAM species, Dittrich et al. (1973) reported data giving a mean activity for RudP carboxylase of 3.3 μmol min^{-1} mg chl^{-1}.

Badger et al. (1975) studied RudP carboxylase and its associated oxygenase activity,

$$\text{RudP} + O_2 \rightarrow \text{PGA} + \text{P-Glycolate} \qquad (4.11)$$

in *Kalanchoe daigremontiana* and found the high affinity form of spinach recently characterized by Bahr and Jensen (1974). Badger's data do not indicate any significant differences between *Kalanchoe* RudP carboxylase/oxygenase and C_3- or C_4-species carboxylase. They obtained a maximum rate of CO_2 fixation of 3.5 μmol min^{-1} mg chl^{-1} and a rate of 0.83 for O_2. The Km (CO_2) was 18 μM, and for O_2 it was 830 μM with Km (O_2) for the carboxylase of 490 μM.

4.2.1.9 Phosphofructokinase (ATP-D-fructose-6-phosphate-1-phosphotransferase)

$$\text{F6P} + \text{ATP} \rightarrow \text{FDP} + \text{ADP}. \qquad (4.12)$$

Phosphofructokinase (PFK) is believed to be a regulatory enzyme in plants (Kelly and Turner, 1969), being inhibited by low concentrations of PEP. Because of this and the modification of the reaction kinetics by citrate, ATP, and some inorganic ions, regulation of glycolysis is possible at the PFK step. In pea seeds, PFK is nearly completely inhibited by PEP concentrations as low as 2–4 μM. This high sensitivity to PEP is also true for PFK from *Atriplex spongiosa*, but mM concentrations are required for inhibition in *Kalanchoe daigremontiana* (Sutton, 1975a–c). The latter observation suggests glycolysis to produce PEP can proceed in *K. daigremontiana* at high concentrations of PEP consistent with the notion that PEP comes from glycolysis in CAM metabolism. In this regard, Sutton (1975a–c) reported that PEP (2 mM) also inhibited *K. daigremontiana* phosphorylase as did glucose (4 mM) and Pi (91 mM).

4.2.1.10 Phosphorylase (α-1.4-glucan: Orthophosphate Glucosyltransferase)

$$(\alpha\text{-D-Glu})_n + \text{Pi} \rightleftarrows \text{G-1-P} + (\alpha\text{-D-Glu})_{n-1}. \qquad (4.13)$$

Phosphorylase is assumed to be involved in starch mobilization during the glycolytic genesis of PEP (see Chap. 3.1.5). The enzyme of *K. daigremontiana* was studied in detail by Sutton (1974a, b; 1975a–c). The pH optimum is about pH 6.0 and the Km for 6-1-P was estimated to be 0.11 mM. There was strong inhibition with 1 mM Pi. Glucose (4 mM) and PEP (2 mM) inhibited to a lesser degree. The data of Sutton show that the *K. daigremontiana* phosphorylase has lower sensitivity to metabolites than the phosphorylase from non-CAM plant species.

4.2.2 The Compartmentation of CAM Enzymes and Metabolites

4.2.2.1 Enzymes

P-enolpyruvate Carboxylase. P-enolpyruvate carboxylase activity was first reported by Bandurski and Greiner (1953) to be in extracts of spinach leaves. Later investigators, e.g., Mazelis and Vennesland (1957), Rosenberg et al. (1958), and

Table 4.6. Distribution of NADP-triose phosphate dehydrogenase and P-enolpyruvate carboxylase in nonaqueous chloroplast preparations[a]

Fraction	Chlorophyll[b]	NADP-TPDH		PEPC	
		N[c]	R[d]	N[c]	R[d]
1	0.05	0.23	1.0	1.5	1.0
2	0.25	0.23	1.0	0.6	0.4
3	0.10	0.20	0.9	0.25	0.17

[a] Data of Mukerji and Ting (1968b).
[b] mg per fraction.
[c] N = Activity in μmol min^{-1} chl^{-1}.
[d] R = Ratio of N to Chl.

Johnson and Bruff (1967) reported activity in chloroplast fractions isolated from spinach. Garnier-Dardart (1965) reported that P-enolpyruvate carboxylase was localized in the chloroplasts of a CAM plant, *Kalanchoe daigremontiana*. Slack et al. (1969) suggested that in maize P-enolpyruvate carboxylase may be bound to the envelope of mesophyll chloroplasts. In our own work, we observed that P-enolpyruvate carboxylase was associated with nonaqueously prepared chloroplasts in *Opuntia*, but not with aqueously prepared chloroplasts (Mukerji and Ting, 1968b). Brandon (1967) reported P-enolpyruvate carboxylase localization in mitochondrial fractions of *Bryophyllum*, and Mazelis and Vennesland (1957) reported it in mitochondrial fractions of cauliflower and pea seedlings. In corn root sections and in *Opuntia* green tissue, we could not obtain evidence for mitochondrial compartmentation (Danner and Ting, 1967; Mukerji and Ting, 1971). These published data taken together seem to suggest some particulate localization of P-enolpyruvate carboxylase.

In nonaqueously prepared chloroplasts from the succulent, *Opuntia ficus-indica*, NADP triose phosphate dehydrogenase maintained almost a constant ratio of activity to chlorophyll (Table 4.6), yet P-enolpyruvate carboxylase was not associated with chlorophyll on a constant basis.

These data suggest that P-enolpyruvate carboxylase may not be associated with chloroplasts, and that it is not localized in the same subcellular compartment with NADP-triose phosphate dehydrogenase. Linear, nonaqueous gradient (CCl_4-hexane) purification of chloroplasts indicate a particulate, but not a chloroplast localization of PEP carboxylase (Fig. 4.4). We conclude from these data that P-enolpyruvate carboxylase may be membrane-associated in situ, but the majority of the activity is not associated with chloroplasts, mitochondria, or microbodies. This conclusion is completely consistent with the metabolic compartmentation data.

Malate Dehydrogenase. Most green plant tissues investigated appear to have three malate dehydrogenase loci; two particulate loci, mitochondrial and microbody; and a soluble or nonparticulate locus. Multiple forms can and do exist within a locus group (Mukerji and Ting, 1968).

In addition, succulent plants have a chloroplast-specific NADP malate dehydrogenase analogous to the NADP form reported in C_3 and C_4 plants (Hatch and

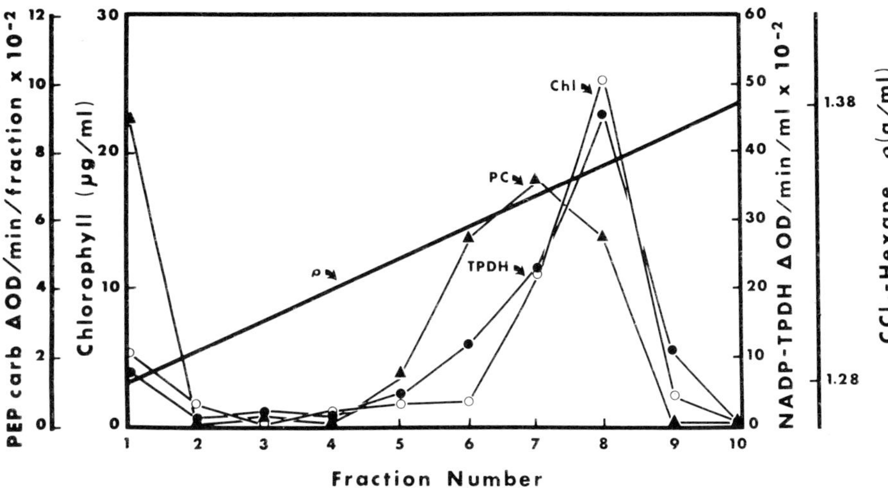

Fig. 4.4. Purification of nonaqueously prepared chloroplasts on a linear carbontetrachloride-hexane gradient (*p*). Whereas the NADP-triose phosphate dehydrogenase (*TPDH*) activity follows chlorophyll (*Chl*) [chloroplasts] exactly, PEP carboxylase (*PC*) is displaced to a lower density indicating absence of chloroplast association (data of Mukerji and Ting, previously unpublished)

Slack, 1969; Johnson and Hatch, 1970; and Ting and Rocha, 1971). This protein is believed to function in part by reduction of oxalacetate to malate during C_4 photosynthesis, and participates in carbon shuttles with the other MDH proteins occurring among compartments.

Because of the fairly conclusive evidence that malate synthesized by dark CO_2 fixation is not in equilibrium with malate synthesized in the mitochondria, it seems reasonable to conclude that the soluble malate dehydrogenase functions to a large extent in the dark CO_2 fixation pathway. We also would predict, that insofar as P-enolpyruvate carboxylase-mediated synthesis of malate takes place in chloroplasts, the NADP malate dehydrogenase would be the coupling enzyme.

Malate Enzyme. Malate enzyme, as well as the other enzymes of the pathway, is known to exist in multiple forms. In apple fruit, Dilley (1966) isolated and purified a form which ran as a single band on acrylamide gels. In corn root tissue, we were able to locate a single isoenzyme (by starch gel electrophoresis) and reported evidence that it was a soluble or nonparticulate protein (Danner and Ting, 1967). Hence, it was concluded that dark CO_2 metabolism, insofar as it is mediated by this particular malate enzyme isoenzyme, is in the cytosol.

In green tissue of *Bryophyllum*, Brandon (1967) reported malate enzyme activity localized in mitochondria. In green cactus tissue, we were able to demonstrate three different proteins with malate enzyme activity apparently localized in three different subcellular compartments; viz., cytosol, mitochondria, and chloroplast fractions (Mukerji and Ting, 1968).

Aspartate Aminotransferase. Aspartate aminotransferase is classically considered to be a mitochondrial enzyme. In higher plant tissues, multiple molecular forms are known (Fasella et al., 1966). *Kalanchoe* leaf tissue has at least two isozymes which are separable on DEAE-cellulose and by starch gel electrophoresis (unpublished). Organelle purification on sucrose gradients has indicated that at least one of these isozymes is located in purified mitochondria, and one other is most likely soluble.

P-enolpyruvate Carboxykinase. Little information is available on the localization of PEP carboxykinase in CAM plants, but Dittrich et al. (1973) obtain activity in both soluble and particulate fractions from pineapple. Classically, the enzyme is considered to be mitochondrial.

4.2.2.2 Metabolites

It is well known that there are multiple compartments of organic acids within the cell (MacLennan et al., 1963; Lips and Beevers, 1966a, b; Osmond and Laties, 1969). In *Bryophyllum* leaves MacLennan et al. (1963) reported data showing that less than 30% of the malate synthesized from acetate was in an active turnover pool and less than 50% of the evolved CO_2 came from citric acid cycle reactions. Lips and Beevers (1966a), in a similar study with corn roots, showed that the malate synthesized from CO_2 was in a pool different from that synthesized from acetate. Evidence for a cytoplasmic pool of malate was presented by Lips and Beevers (1966b) and by Osmond and Laties (1969). In the later study, data were obtained which indicated that malate synthesized from CO_2 fixation in beet discs was present in a cytoplasmic pool with a relatively short half-life and a vacuolar pool with a longer half-life. Similar results were obtained by Kluge and Heininger (1973) with *Kalanchoe daigremontiana* leaves. In work with *Opuntia*, other succulents, and *Opuntia* root tips (Ting and Dugger, 1966; Ting and Dugger, 1968; Ting and Zschoche, 1970) evidence was found for a two-phase decarboxylation of malate from CO_2. These data were taken to indicate two pools of organic acids (Ting and Dugger, 1968). Buser and Matile (1977) found more than 90% of malic acid of *Kalanchoe* leaf cells to be localized in the vacuoles.

As previously indicated, during dark fixation of CO_2, no radioactive carbohydrate products are formed; only organic and amino acids are significantly labeled. In cactus root tips, after a 2-h dark fixation period, 80% of the ^{14}C was in organic acids (Ting and Dugger, 1966) with a distribution of 66.4% in malate, 11.4% in citrate, and trace amounts in fumarate, succinate, and isocitrate. Because the tissue was actively respiring (about 200 μl g $fr.wt^{-1}$ h^{-1}), the data were taken to indicate compartmentation of the products outside the mitochondria.

These data indicate that malate and subsequent products of dark CO_2 fixation are not in general equilibrium with other metabolic processes within the cell, and furthermore, the data do not suggest a lack of glycolytic or citric acid cycle activity.

Unfortunately, practically nothing is known about the compartmentation of CAM metabolites other than the organic acids. Of course, starch can easily be recognized as a constituent of chloroplasts (see Fig. 2.8). Further investigations to clarify compartmentation of metabolites are desirable to solve the problem of cellular control of CAM.

4.2.2.3 Conclusions

All the data in the literature taken together strongly suggest that the products of dark CO_2 fixation are compartmented and largely separated from mitochondrial, microbody, and probably chloroplast metabolism. It is generally suggested that dark CO_2 metabolism is localized in the cytosol. The main enzymes mediating the reactions, viz., P-enolpyruvate carboxylase, malate dehydrogenase, malate enzyme, and aspartate aminotransferase, all exist as isozymes or in multiple molecular forms. It is clearly shown that these multiple molecular forms are localized in different subcellular compartments. If dark CO_2 fixation metabolism is composed of cytosol reactions, then one expects that the soluble isozymes mediate the flow of carbon. The other isozymes would participate only to the extent that the isozyme reactions are in equilibrium and that carbon flows from one compartment to another.

4.2.3 Models of Metabolic CAM Control

4.2.3.1 Control of CAM During the Dark Period

The capacity of CAM plants to accumulate malic acid at night has been interpreted by Sutton (1975a–c) to be a consequence of the special regulatory properties of the CAM plant glycolytic pathway in comparison to plants without CAM. As outlined in Chapter 4.2.1, differences exist in the regulatory behavior of phosphorylase and phosphofructokinase. Inorganic phosphate, glucose, and PEP are inhibitory to phosphorylase from the CAM plant *Kalanchoe daigremontiana*, but to a lesser extent than the enzyme obtained from the C_4 plant *Atriplex spongiosa*. Phosphofructokinase from the CAM plant is inhibited by malate. PEP, however, is a strong inhibitor of phosphofructokinase only in the non-CAM plant. These characteristics of the two important enzymes of the glycolytic pathway allow CAM plants substantial amounts of PEP for β-carboxylation even during accumulation of substantial concentrations of malate.

Pool sizes of glycolytic pathway intermediates (G-6-P, F-6-P, FDP, pyruvate) in the dark and with and without external CO_2 have been measured by Cockburn and McAulay (1977). They found an increase of PEP under CO_2-free air. Sutton's results indicate a high PEP level could inhibit phosphorylase, and in fact Kluge (1969b) found in several CAM plants that starch consumption during the night was drastically inhibited with CO_2-free air.

From the data of Cockburn and McAulay (1977) it appears that internal CO_2 concentrations are part of the circuit (see Chapter 4.6) regulating carbohydrate mobilization and malic acid synthesis: Briefly after light/dark transitions, internal CO_2 concentration rises because of respiration. High CO_2 activates the glycolytic breakdown of carbohydrate from glucan to G-6-P in a manner as yet unknown. Simultaneously, PEP-C is inhibited by high CO_2. Thus, the G-6-P would activate PEP carboxylase favoring malic acid synthesis and keeping internal CO_2 at a level low enough to allow uninhibited action of PEP-C. Maximum PEP-C activity reduces the concentration of PEP which results in increased phosphofructokinase activity. Finally, malic acid accumulation reaches saturation. In this phase, the

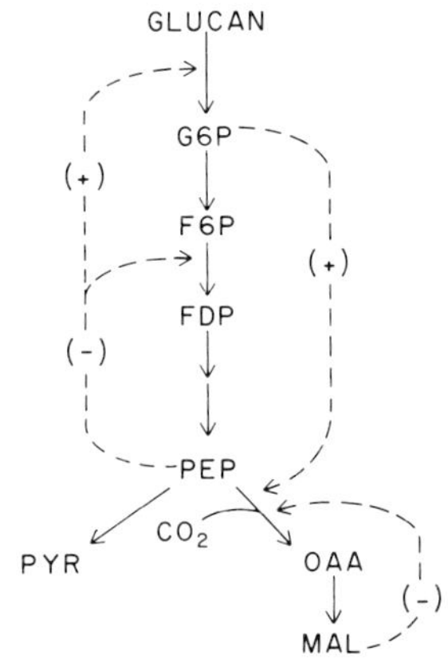

Fig. 4.5. Proposed substrate level modulation of carbon flow from glucose to malate in CAM. (+) and (−) = positive and negative modulation respectively

increasing level of malate is expected to inhibit PEP-C. Thus, the internal CO_2 concentration will rise and also inhibit PEP-C. Reduced PEP-C activity causes inhibition of phosphofructokinase.

The above hypothesis of Cockburn and McAulay (1977) does not include the vacuoles as the store for malic acid. However, there is increasing evidence that the storage behavior of the vacuole for malic acid plays a dominant role in the regulation of the cytoplasmic malic acid pool (Kluge and Heininger, 1973; von Willert and Kluge, 1973; Lüttge and Ball, 1974; Lüttge et al., 1975, 1977; Kluge, 1977a; see Chap. 4.2.3.2,e). Kluge (1977a) showed that during most of the dark period, the vacuole acts as a strong sink for malate and that at the end of the dark period the storage capacity of the vacuole for malate becomes exhausted. Thus, in the light malic acid would be enriched in the cytoplasm and inhibit further malate synthesis by feed-back inhibition of PEP carboxylase. The question of malic acid storage is also discussed in Chapters 3.1.7 and 4.2.3.2,e of this book.

Our understanding of the regulation of carbon flow during acidification (or deacidification) comes largely through studies of metabolite compartmentation and enzymic properties. The strict compartmentation of the organic acid products of CAM malate, citrate, and other acids, must mean that there is a degree of control at the branch point between pyruvate kinase (PEP→PYR) and PEP carboxylase (Fig. 4.5). The nature of this regulation which keeps citric acid cycle and CO_2 fixation products separate is virtually unknown, but must somehow be related to the multiple forms of common enzymes, e.g., malate dehydrogenase.

The carbon flow from glucose to malate could be regulated at several steps (Fig. 4.5). Enzymes such as phosphorylase, phosphofructokinase, and PEP carboxylase are affected by their own products or substrates. Phosphorylase producing phosphorylated hexose from glucose is inhibited by PEP, glucose, and phosphorus. Further, phosphofructokinase is inhibited by PEP, and PEP carboxylase is activated by glucose-6-phosphate and inhibited by malate. Hence, hydrolysis of glucan to form glucose-6-phosphate will tend to activate PEP carboxylase until malate builds up, when it will reduce further carboxylation. If this happens, PEP will increase, thereby inhibiting both the production and utilization of glucose-6-phosphate.

The regulation of the branch of PEP producing either pyruvate or oxalacetate is poorly understood, but ATP and/or ADP could be a factor (Ting and Osmond, 1973c).

4.2.3.2 Control of CAM During the Dark/Light Transient and During the Light Period

a) Control by Thermoperiod

Brandon (1967) postulated that diurnally changing thermoperiods must be responsible for the control of CAM. This hypothesis is based on in vitro experiments which showed that the malate-synthesizing enzymes (PEP carboxylase plus MDH) reach their optimum at 35° C, the thermal optimum of "malate enzyme" which may initiate malic acid consumption during CAM, was not reached even at 53° C. Thus, with in vitro combinations of PEP-C, MDH, and malate enzyme, net malate synthesis was observed only at temperatures clearly below 20° C. Above that temperature, depletion of malate dominated. Brandon (1967) concluded from these results that functioning of CAM predominates at low night and high day temperatures.

However, there is ample evidence that CAM can also proceed under constant temperature regimes. For example, many results published by Kluge's group (see Kluge and Fischer, 1967) were obtained with constant temperature during the day/night cycle.

CAM can also occur under tropical conditions with very small diurnal temperature fluctuations, for example in succulent orchids growing at Singapore (Avadhani, personal communication). In a seasonal study of *Opuntia basilaris*, CAM correlated with water availability and not with thermoperiod (Szarek and Ting, 1974a). Diurnal acid fluctuations were as great during the summer as during the winter, providing water was available (Ting, 1976b). *Dudleya farinosa* performed CAM in its natural stands along coastal California where temperatures during day/night are very constant (Bartholomew, 1975; see Fig. 6.6). These arguments suggest strongly that diurnal fluctuations of temperature cannot be generally a major controlling mechanism of CAM regulation. However, it will not be denied that temperature regimes may substantially modulate performance of CAM insofar as the amplitude of malic acid oscillation is increased by favorable diurnal fluctuations in temperature (see Wolf, 1960; Ranson and Thomas, 1960).

b) Control by Endogenous Rhythms of Enzyme Activities

Control of CAM has been interpreted as a manifestation of a biological clock (Queiroz, 1974, 1975). This endogenous clock is suggested to control diurnal cycles of activities of malic acid-synthesizing and malic acid-consuming enzymes. The data provided by Queiroz and co-workers show PEP-C oscillating diurnally with high activity at night and low activity during the day (Queiroz et al., 1971). The activity of malic enzyme oscillates out of phase with respect to PEP-C (Fig. 4.6).

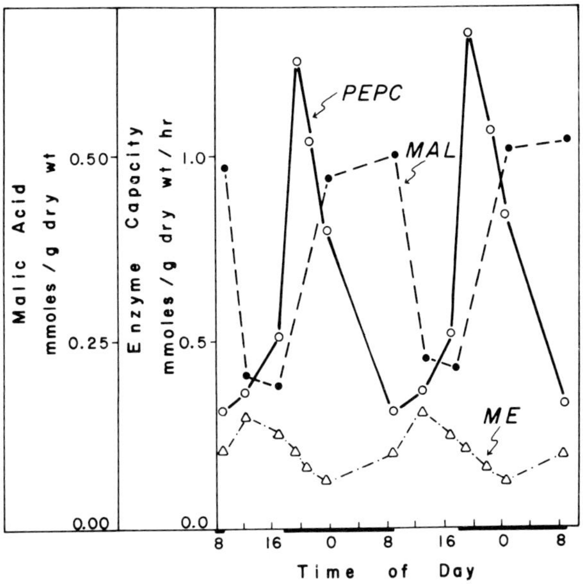

Fig. 4.6. Circadian oscillation of PEP carboxylase, malate enzyme and malic acid in *Kalanchoe* over a 48 h period (data of Queiroz and Morel, 1974, by permission)

Arguments supporting the biological clock model come from the finding that the activity of certain enzymes not directly integrated in CAM (for example aminotransferase) also oscillates diurnally. According to Queiroz (1974), this suggests the action of a basic oscillator ("Zeitgeber") of yet unknown nature which controls a variety of enzymes, including PEP-C. The above hypothesis would also explain the finding of Queiroz et al. (1972) that the phase of the diurnal oscillations in PEP-C and aminotransferase can be shifted parallelly under certain experimental treatment.

There is no doubt that features of CAM other than oscillating enzyme activities may exhibit characteristics of endogenous rhythms. For instance, the typical CO_2 exchange pattern of CAM proceeds for some time in continuous light or dark (for detailed discussion of this phenomena see Chap. 5.1.5). In addition, acidification and deacidification continue for a while under constant light (Kluge, 1969c; Lüttge and Ball, 1977).

However, the weak point in the above regulation model still remains the yet unknown nature of the basic oscillator.

c) Control by Competition

A further model (Beevers et al., 1966) is founded on the assumption that in CAM the PEP-C pathway of CO_2 fixation (i.e., malic acid synthesis) and photosynthetic CO_2 fixation compete for CO_2 as a common substrate. Hence, during the day, malic acid synthesis should be retarded because, due to competition, the CO_2 as the substrate of PEP-C will be in short supply.

With malic acid synthesis being inhibited, malic acid consumption dominates, allowing net consumption of malic acid ("deacidification"). During the night, malic acid may accumulate since photosynthesis can no longer compete for CO_2, i.e., PEP-C would operate at its full capacity.

In fact, leaves of CAM plants may perform net malic acid synthesis even during the day at abnormal high concentrations of CO_2 in the atmosphere (Bruinsma, 1958). This suggests that now the concentration of the CO_2 is high enough to avoid competition because both PEP-C and RudP-C are sufficiently supplied with their substrate.

However, several findings are inconsistent with the above model. Firstly, malic acid accumulation occurs even at normal CO_2 concentrations during irregular light periods given during the time of the normal dark period. This suggests that simple competition for CO_2 is not the only factor involved in regulation of the diurnal CAM cycle. Secondly, the Km of PEP-C for CO_2 (0.02 mM; Ting and Osmond, 1973a–c) is at least as low as the Km (CO_2) of RudP-C (0.019 mM, see Chap. 4.2.1). Furthermore, PEP-C activity in CAM plants is doubtless higher than that of RudP-C (Kluge and Osmond, 1972; Osmond, 1976a). With the background of these data it is difficult to postulate an effective competition for CO_2 between PEP-C and RudP.

A variant of the "competition model" has been postulated by Osmond and Allaway (1974). These authors assume competition between PEP-C and RudP-C pathways for C_3 carbon compounds rather than by CO_2.

According to Osmond and Allaway (1973), PEP is generated during the day at the expense of 3-PGA from the RudP-C reaction. However, 3-PGA is preferably consumed in order to maintain action of the reductive pentose phosphate cycle, thus allowing only limited production of PEP. The limited supply of its substrate would inhibit PEP-C.

This version of the competition model avoids the inconsistencies arising from the Km values and activity ratios of both carboxylases as outlined above. However, malic acid accumulation occurring in the light during the time of the regular dark period cannot be explained.

d) Regulation by Feedback Inhibition

As outlined in Chapter 3.2, CO_2 is an essential intermediate in carbon flow of CAM on the way from malic acid to carbohydrate. Furthermore, it has been argued above that Km values and activities of PEP-C and RudP-C as estimated in CAM plants should favor PEP-C in case of free competition for CO_2 between both enzymes. Any relevant model has to explain, against the background of the above data, how CO_2 produced by malic acid decarboxylation can enter the Calvin cycle nearly quantitatively, thus providing carbohydrates, rather than being recycled into malic acid by PEP-C.

Effective carbon flow from malic acid to carbohydrate with CO_2 as an intermediate could be explained by the postulate that PEP-C might be temporarily suppressed as a competitor for CO_2, thus interference with RuDP-C would be avoided. Both, the biological clock model and the competition model (see above) account for this postulate. However, considering the obvious limitations of these hypotheses, Kluge (1969b), Kluge and Osmond (1972) have suggested a model which supposes that PEP-C might be controlled by a feedback mechanism based on a direct inhibition of the enzyme by malate (see also Chap. 4.2.1.3,a). This model was stimulated by results of Ting (1968a) who showed that PEP-C of maize roots is inhibited by malate and oxalacetate. Also PEP-C obtained from CAM plants is inhibited by malate (Queiroz, 1967; Kluge and Osmond, 1971), and by pyruvate (Kluge and Osmond, 1971).

The original feedback inhibition model suggested that as long as the cellular malic acid level is high enough (i.e., during the early hours of the day), PEP-C would be inhibited and fixation of (endogenous) CO_2 via the reductive pentose phosphate cycle would be facilitated. With the malic acid level being low again (i.e., during the late hours of the light period and during nocturnal acidification), PEP-C would no longer be inhibited, thus allowing malic acid synthesis.

The feed-back model is supported, apart from the above enzymological data, by in vivo experiments (Kluge, 1969b), where detached phyllodia of *Kalanchoe tubiflora* were fed with $^{14}CO_2$ under illumination at the beginning (malic acid level being high) and again at the end of the light period (malic acid level being low). In the early light period, the label was incorporated preferentially into sucrose rather than into malate. This indicates that CO_2 fixation proceeded mainly via the reductive pentose phosphate cycle. However, in the identical experiment performed at the end of the day, the bulk of the label was found in malic acid, suggesting PEP-C was no longer inhibited and competed successfully for CO_2. Avadhani et al. (1971) reported similar observations obtained with *Sedum praealtum*.

e) Regulation by Feedback Inhibition of PEP-C and Transport Control of Malic Acid at the Tonoplast

The "feed-back inhibition model," as outlined above, must be considered as too simplified, because it does not account for the fact that PEP-C is mainly localized in the cytosol (see also Chap. 4.2.2.1) and most of the malic acid is stored in the vacuoles. However, logically only the malic acid present in the cytoplasm may directly affect PEP-C. Hence, postulating control of PEP-C activity by malic acid demands interrelationships between the vacuolar and cytoplasmic malic acid pools. Consequently, Kluge and Heininger (1973) concluded that fluxes of malic acid determine substantially the malic acid level of the cytoplasm which finally controls PEP-C activity.

Attempting to clarify properties of malic acid transport across the tonoplast, Lüttge and Ball (1974; Lüttge et al., 1977) have studied leaf slices of *Kalanchoe daigremontiana* suspended in solutions of varied osmotic potential.

Their data showed that the loss of malic acid from the vacuole depended strongly on the water potential of the external medium which varied due to different concentrations of sorbitol. Substantial efflux of malic acid from the vacuole was achieved at high water potential of the medium, and the export of malic acid decreased if the water potential of the medium was lowered. Malic efflux from the vacuole was lowest when the vacuolar content and the external solution were isotonic.

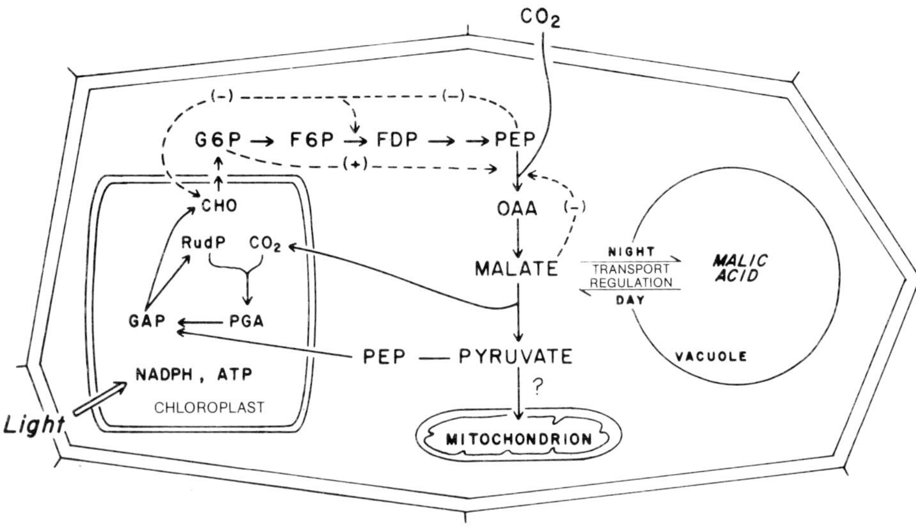

Fig. 4.7. Conceptual model for the regulation of CAM metabolism. The main features are the substrate level modulation of glycosis and dark CO_2 fixation at night, the bidirectional cytoplasm—vacuole transport regulation night and day, and the photosynthetic assimilation of CO_2 from malate decarboxylation

These results have been explained by means of a turgor mechanism involved in control of malic acid transport across the tonoplast (Lüttge and Ball, 1974; Kluge and Lüttge, 1974; Lüttge et al., 1975, 1977). The basic assumption is that increased turgor enhances the net efflux of malic acid from the vacuole into the cytoplasm, and vice versa. According to this idea, the above experimental findings might be interpreted as follows:

High water potential of the external solution will facilitate osmotic water uptake by the cells, i.e., their turgor will increase. In vivo, an increase in turgor could be achieved by malic acid accumulation in the vacuole and concomitant osmotically driven influx of water into the vacuole.

Furthermore, Lüttge et al. (1975) showed malic acid synthesis by dark CO_2 fixation and malic acid accumulation to occur in leaf slices of *Kalanchoe daigremontiana*, provided the cells and external medium were approximately isotonic (i.e., when the turgor pressure of the cells was low). Moreover, the tissue slices "acidified" in a medium of proper water potential or "deacidified" if transferred into a medium of higher water potential. Hence, a rhythmic change in malic acid content of the cells could be achieved by varying the water potential of the external solution (i.e., indirectly the turgor of the cells).

The above model (see Fig. 4.7) explains CAM in terms of a tension-relaxation mechanism (see Lüttge et al., 1975). Hence, it is consistent with the general principles of rhythmic phenomena in plants which appear to have endogenous oscillations (Bünning, 1973; see also Chap. 5.1.5). Further, it can explain CAM under constant photo- and thermoperiods consistent with observations of CAM in constant environments.

However, final support of the above model demands direct measurements of turgor in CAM-performing cells. First attempts in this direction were provided by Lüttge and Ball (1977) who measured in intact leaves of *Kalanchoe daigremontiana* both total water potential and osmotic and osmotic potential, and calculated the pressure potential. They found that the malic acid rhythm of CAM contributed to a rhythm of water potential. However, the pressure potential (i.e., the turgor) remained practically unchanged during the diurnal cycle. This, however, would be inconsistent with the as asumption of a turgor mechanism being involved as a trigger in regulation of CAM in vivo. As indicated elsewhere, the hypothesis is tentative and demands further investigation.

The exact role of light in the regulation of CAM phenomena is not at all clear, and not an integral part of the above model. Whether light is simply involved in the continued photosynthetic consumption of malate, whether it modulates enzymes involved in deacidification, perhaps even altering tonoplast permeability directly (Wolf, 1960; Kluge, 1971 b; Nalborczyk et al., 1975) is not understood at this time.

Some of the recent assumptions dealing with the metabolic control of CAM are summarized in Figure 4.7.

4.3 Modification of the Diurnal Malic Acid Cycle by External Factors

4.3.1 Effects of Temperature

Effects of temperature on the diurnal pattern of CO_2 exchange in CAM are discussed in Chapter 5. In the following section, effects of temperature on the day/night acid cycle of CAM will be briefly considered (for detailed phenomenology see Wolf, 1960).

It must be mentioned that the discussion of temperature effects is complicated by the fact that CAM is modified both by direct temperatures and to a great extent by temperature pretreatment. In general, it is well known that low night temperatures favor nocturnal acid accumulation (for example De Vries, 1884; Richards, 1915). Below 5° and above 30° C, nocturnal malic acid synthesis is inhibited.

Malic acid consumption of the light period is accelerated by high temperatures (see Wolf, 1960; Kluge, 1968 b; Moradshahi et al., 1977; Somers, 1951). The reason for stimulation of deacidification by increased temperatures is not clear. With high temperatures, respiration will be enhanced, and much malic acid may be consumed by respiration as well as by decarboxylation. Furthermore, temperature-dependent alterations in tonoplast permeability may facilitate malic acid losses from the vacuole.

4.3.2 Effects of Light

Malate consumption during the day is accelerated by increased light intensities (Kluge, 1968 b; Brunnhöfer et al., 1968 a, b; Osmond, 1978; see Fig. 5.10). Also, the higher the light intensity during the deacidification phase, the lower the minimum value of malic acid reached (cf. Kluge, 1968 b and Fig. 5.10). Similarly to temperature, daytime light intensity will affect the acidification rate during the subsequent dark period. Generally, the difference between night-time maximum and the day-time minimum malic acid is increased after high light intensity treatment. Net dark CO_2 fixation is also increased with high light intensity during the previous light period. The greater rates of dark CO_2 fixation and increased malic acid production after high light intensity are probably results of more carbohydrate biosynthesis in the light, and hence more available PEP, plus the greater vacuole storage capacity for malic acid after the daytime depletion.

Aside from the above effects, and from the fact that light results in the dramatic shift from malic acid synthesis to malic acid consumption and concomitant production of photosynthetic products, there is little direct evidence of light influencing CAM metabolism at the metabolic level (see Chap. 4.2.3.2,e). To account for the shift from acidification to deacidification, Kluge (1971 b) proposed a light-dependent change in tonoplast permeability allowing stored malic acid to flow into the cytosol, inhibit PEP carboxylase, and be decarboxylated. Garnier-Dardart and Queiroz (1974) proposed a light activation of malate enzyme similarly to pyruvate Pi dikinase (Hatch and Slack, 1969) and NADP-malate dehydrogenase (Hatch, 1976).

The diurnal fluctuation of CAM enzymes and the alteration of metabolism because of short days so extensively studied by Queiroz and co-workers (see Queiroz, 1974), is another example of light alteration of CAM. There are indications that the latter might be at least partially a phytochrome response (Queiroz, 1972, 1974; see also Narlborczyk et al., 1975).

4.3.3 Effects of Ions

It is generally known that NaCl tends to decrease the ratio of malate to aspartate formed during dark CO_2 fixation. Marine algae and angiosperms tend to form more aspartate than malate (Joshi et al., 1962). Webb and Burley (1965) studied facultative and obligate halophytic succulent chenopods. In these halophytes, NaCl was associated with an increased ratio of amino acids to organic acids (Table 4.7). There is no evidence, however, that these succulents of the genera *Batis*, *Salicornia*, and *Borrichia* are CAM plants.

Winter et al. (1974), studying *Mesembryanthemum crystallinum*, showed that the NaCl-enhanced dark CO_2 fixation was accompanied by relatively more organic acid synthesis, and not amino acid synthesis. Hence, this aizoid differs from the halophytic chenopods studied by Webb and Burley (1965).

NaCl has a strong influence on the activity of the important CAM enzyme, PEP carboxylase (Table 4.5) and malate dehydrogenase. At low NaCl concentrations or ionic strength, PEP carboxylase and malate dehydrogenase may be activated, while at high concentrations severe inhibition occurs (Osmond and Greenway, 1972; Ting and Osmond, 1973 a, b; Weinberg, 1967; Zschoche and Ting, 1973; von Willert, 1975 a).

In the case of purified microbody malate dehydrogenase, this is strongly pH-dependent with marked activation at pH 6.0 at low ionic strength (Zschoche and Ting, 1973). Malate enzyme of *Opuntia* is also inhibited by NaCl (Mukerji and

Table 4.7. Percentage distribution of recovered C^{14} after dark fixation of $C^{14}O_2$ by *Spartina alterniflora* in presence and absence of NaCl[a]

Time in hours	Amino acids		Organic acids	
	−NaCl	+NaCl	−NaCl	+NaCl
0.5	59.8	66.8	40.2	30.4
1	41	59.4	59	36.7
1.5	37.4	56	62	41
3	11.4	46.5	88.6	50.5
4.5	—	37	—	60

[a] From Webb and Burley, Can. J. Bot. *43*, 281–285 (1965).

Ting, 1971). Winter and von Willert (1972), Kramer and von Willert (1972) reported that *Mesembryanthemum crystallinum* could be induced to take up CO_2 in the dark and accumulate malic acid if irrigated with NaCl (see Fig. 5.8). The increased malic acid accumulation was accompanied by an increase in PEP carboxylase. The physiological basis for CAM induction is not entirely clear, but may be related to unfavorable water potentials accompanying the NaCl treatment (Winter, 1973, 1974; for more detail cf. Chap. 4.3.4.2).

Similarly, *Portulacaria afra*, a succulent termed weak CAM by Neales (1975) is shifted from day-time CO_2 uptake to nocturnal CO_2 fixation and (relatively little) acid fluctuation by NaCl stress. This shift is concomitant with a day-time closure and nocturnal opening of stomata and an increased diurnal fluctuation of organic acids (Fig. 4.8). Since water stress was accompanied by the same shift in malic acid metabolism, the NaCl effect here may be indirect and simply related to physiological drought.

Brownell and Crossland (1974) found that plants having C_4 photosynthesis require Na^+ as a micronutrient. Furthermore, these authors detected that the CAM plant *Kalanchoe daigremontiana* showed a Na^+ requirement for growth, however, only under environmental conditions in which the plants harvest external CO_2 entirely during the night. In contrast, those plants growing on long-day conditions and high irradiance (i.e., with substantial CO_2 fixation via the C_3 mode of photosynthesis) did not use Na^+ as a micronutrient. These results suggest close relationships between C_4 photosynthesis and CAM. There is not yet a convincing causal explanation for the Na^+ requirement of the C_4 pathway of CO_2 fixation.

Lai (1977) studying the effect of NaCl on CO_2 exchange by *Kalanchoe daigremontiana* observed a shift from daytime to nighttime CO_2 influx.

As indicated earlier (Chap. 4.2.1.1), inorganic phosphate (Pi) lowers the inhibition of PEP carboxylase by malate and NaCl in in vitro systems (von Willert, 1975b, c). Von Willert found that in *Mesembryanthemum crystallinum* the induction of CAM during irrigation with NaCl solutions is accompanied with an increase in the Pi level of the leaves. Therefore, it was argued that inorganic phosphate protected PEP carboxylase from the inhibition of NaCl and malic acid. Thus, CAM could operate in salt-stressed *Mesembryanthemum* plants in spite of increasing NaCl and malic acid levels in the cells. It should be noted, however,

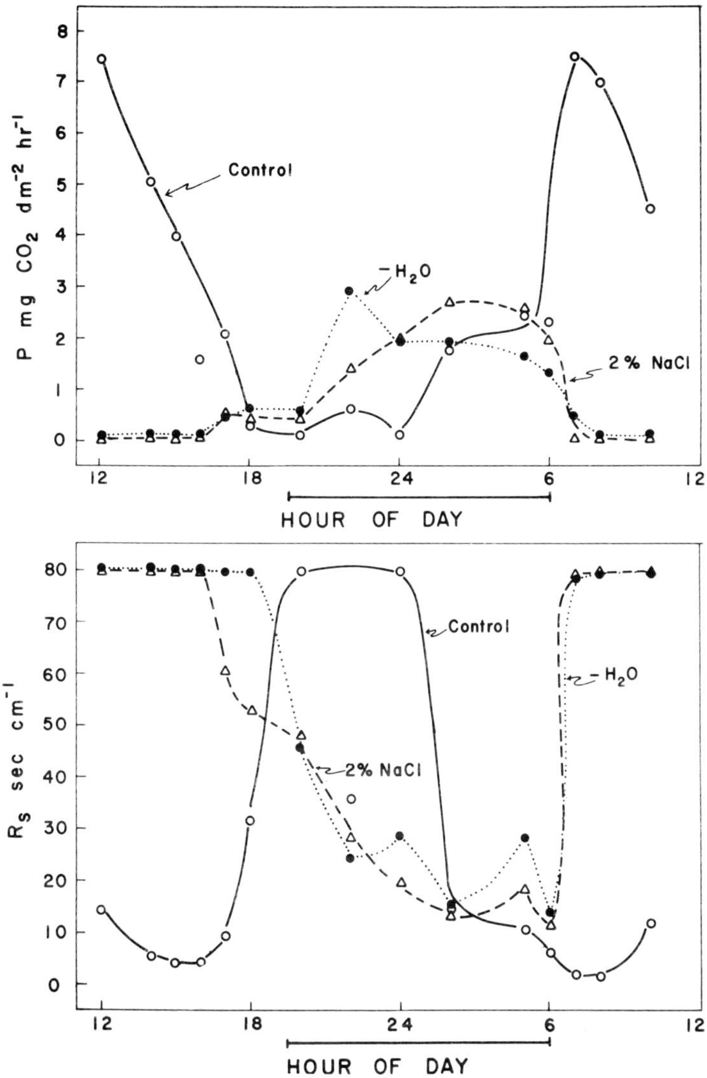

Fig. 4.8. Metabolic shift in *Portulacaria afra* accompanying water or NaCl stress. Control plants were watered frequently to prevent stress, whereas $-H_2O$ were watered weekly and NaCl plants were irrigated with 2% NaCl. *P*, CO_2 uptake; R_s, leaf resistance to gas transfer (data of Ting and Hanscom, 1977)

that the in vitro protective effects of inorganic phosphate on PEP carboxylase required concentrations of phosphate much higher than 30 mM. We do not know if in vivo PEP carboxylase is exposed to such high levels of phosphate.

4.3.4 Effects of Water Relations and the Question of "Facultative" CAM Plants

Effects of water relations on the expression of CO_2 exchange patterns linked to CAM are discussed in detail in Chapter 5. In relatively few studies, CAM

Table 4.8. CO_2 exchange and transpiration in obligate CAM plants in response to drought and its relation to dark acidification, transpiration ratio, and growth rate[a]

Water potential (bars): soil	− 1	− 5	> − 50
plant	− 3	− 6	− 15
Exchange of external CO_2: night	+	+	±
beginning of the light period	+	+	−
end of the light period	+	−	−
Δ malic acid (% maximum)	100	50 –80	25
Transpiration ratio (gH_2O/gCO_2)	50–600	18 –50	zero
Growth rate (g m^{-2} day^{-1})	5– 20	0.5– 1.5	zero

[a] From Osmond, 1978.

features other than gas exchange have been studied in response to water relations. The literature on CAM indicates that two diverging patterns of CAM behavior exist in response to water deficiency, namely (a) modification of CAM in those plants which perform CAM under nearly all circumstances (after Osmond, 1978 "obligate" or "constitutive" CAM plants), and (b) "induction" of CAM by water deficiency in certain plant species which show normal C_3 photosynthesis if well supplied with water ("facultative" or "inducible" CAM plants).

4.3.4.1 Effects of Drought on CAM in "Obligate" CAM Plants

In cactus, there is evidence (Table 4.8, Fig. 6.5) that the amplitude of the diurnal malic acid rhythm becomes smaller if the plant water potential is lowered in response to low soil water potentials (Szarek et al., 1973; Szarek and Ting, 1974b, 1975a, b; Lüttge and Ball, 1977; Osmond, 1978). This is most likely caused by stomatal closure reducing the availability of external CO_2 for dark CO_2 fixation (see also Chap. 5 and 6), and perhaps that the storage capacity of the vacuoles for malic acid might be affected (see Kluge and Heininger, 1973; Lüttge et al., 1975; Lüttge and Ball, 1977). In *Tillandsia usneoides*, net dark fixation is absolutely increased by experimental drought rather than relatively to drought-depressed light CO_2 fixation (Kluge et al., 1973). One would expect that in the latter case nocturnal malic acid synthesis should also be enhanced. Unfortunately, the malic acid rhythm concomitant with the cycle of CO_2 exchange has not been measured in the experiments with *Tillandsia usneoides*. For more details compare Chaps. 5.1.3.2, c; 5.1.4.2, d; 5.3.4.2; 6.2.2.1.

4.3.4.2 Induction of CAM in "Facultative" CAM Plants

The phenomenon of CAM induction in plants which otherwise behave as C_3 plants has been extensively reviewed by Winter and Lüttge (1976). Hence in this book considerations dealing with the problem of CAM induction will be restricted to some basic facts (see also Chaps. 4.3.4; 5.1.3.2, and 6.2.2.3).

Winter and von Willert (1972) were the first to show that certain plant species shift from the C_3 mode of photosynthesis to CAM in response to external factors. Studying effects of NaCl treatment in the facultative halophyte *Mesembryanthemum crystallinum*, the authors found induction of CAM with all its attributes when plants cultivated in salt-free soils were irrigated with NaCl solutions. Na_2SO_4, KCl, and K_2SO_4 had effects on CAM similar on CAM to those of NaCl. Subsequently, induction of CAM by salt treatment was found in other Aizoaceae (Winter, 1973; Treichel, 1975; Treichel and Bauer, 1974), in *Aster tripolium* (Ganzmann and von Willert, 1973), in *Portulacaria afra* (Ting and Hanscom, 1977) and in *Peperomia obtusifolia* (Hanscom and Ting, 1978).

Originally, induction of CAM was believed to be a salinity effect. However, it is now well established that water stress also induces CAM. This has be shown clearly for *Mesembryanthemum crystallinum* (Winter, 1974; Winter and Lüttge, 1976), in *Sedum acre* (Kluge, 1977b), and in *Portulacaria afra* (Ting and Hanscom, 1977; see Fig. 4.8). Hence, it is most likely that the salinity effects on CAM induction result from osmotic effects on low plant water potential. Indeed, CAM was only induced in *Portulacaria* and *Peperomia* in which leaf water potentials were less than −10 bars (Ting and Hanscom, 1977).

It could be shown in laboratory experiments (Winter, 1973; Winter and Lüttge, 1976; Kluge, 1977; Ting and Hanscom, 1977) that CAM induced by stress can be reversed to the C_3 mode of photosynthesis if the stress is withdrawn. In contrast to CAM induction, which usually demands about 5 to 8 days in *M. crystallinum* and *Sedum acre*, the reversal of CAM to the C_3 mode of photosynthesis proceeds only during hours after offset of stress (Winter, 1973; Kluge, 1977; see, however, Ting and Hanscom, 1977). An unequivocal concept for the mechanism of CAM induction is lacking (cf. Winter and Lüttge, 1976). However, there is evidence that CAM induction is accompanied by an increase in total PEP carboxylase activity (Winter, 1974; von Willert, 1975; Treichel et al., 1974; Kluge, 1977). It is therefore reasonable to postulate that induction of CAM may be at least partially controlled at the enzyme level. However, we do not know if CAM in uninduced plants is really limited by PEP carboxylase activity. Nevertheless, the increase in total PEP carboxylase activity during stress seems to be one of the rare examples found in higher plants where actual induction of an enzyme can be observed.

As indicated by kinetic and electrophoretic studies in *Mesembryanthemum crystallinum*, not only total PEP carboxylase activity, but also the properties of the enzyme undergo alterations both during induction of CAM and during its reversion (von Willert, 1975; Treichel et al., 1974; von Willert et al., 1976b). However, more experiments with more species are desired before generalizations. For example, preliminary experiments on *Sedum acre* (Kluge and Knopf, unpublished) provided no indication of alterations in the kinetic or electrophoretic behavior of PEP carboxylase during CAM induction by water stress.

A point where the CAM induction could also be controlled are the stomata themselves. The data seem to suggest that the first response to stress is stomatal closing followed by induction of other parameters of CAM. Careful study of Figure 4.8 shows that *Portulacaria*, when unstressed, have stomata open during the day and again toward the end of the night period. When stressed, stomata close during the day, but still show nearly the same degree of opening at night.

It is well documented that water stress may increase the level of abscissic acid (ABA) in cells (Itai and Benzioni, 1976), and there are reports that application of this hormone to leaves enhances the activity of PEP carboxylase (Huber and

Sankhla, 1976). Hence, it is tempting to speculate that the enhancement of PEP carboxylase activity during CAM induction caused by water stress is mediated by plant hormones, particularly by ABA. Attempts to induce CAM in *M. crystallinum* by application of external ABA to the leaves, to the roots or to leaf slices failed, but ABA caused stomatal closure and initiated fluctuation of organic acids in *Peperomia*. The acid fluctuation did not reach the level of water-stressed plants (see Hanscom and Ting, 1978).

Apart from the many open questions which still exist in connection with CAM induction by salinity and water stress, it can be predicted that further investigations will provide considerable progress on the problem of CAM control and regulation.

4.3.5 Effects of Oxygen and Carbon Dioxide

4.3.5.1 Oxygen

Moyse (1955a, b), Moyse and Jolchine (1956) found that malate does not accumulate in the dark in the absence of O_2. Later experiments of Avadhani (1957) fully supported this observation. Ranson and Thomas in their review (1960) concluded that no appreciable acid accumulation occurred under anaerobic conditions. Because Moyse found dark acidification in O_2-rich, CO_2-deficient air, he concluded that O_2 acts by promoting decarboxylation reactions. This conclusion does not, of course, account for the inhibited malate accumulation in low O_2, CO_2-sufficient air.

Furthermore, inhibitors (of respiration) such as cyanide, azide, and arsenite also inhibit malate synthesis (Brown, 1956; Kinraide and Behan, 1975). Despite the fact that respiration is known to proceed as well at 1% O_2 as at 20%, these studies implicate aerobic respiration as the main oxygen-related factor. Brown (1956), however, showed that malonate reduced oxygen uptake by 50%, yet malate accumulation was largely unaffected.

Ranson and Thomas (1960) discuss oxygen consumption during dark CO_2 fixation and indicate the difficulties of using respiratory quotient data to ascertain the oxygen effect.

Although to some extent we can conclude that normal aerobic metabolism is necessary for CO_2 fixation and acidification, a more exact role for O_2 may be indicated.

By studying the effect of *Bryophyllum* leaf section thickness on respiration, Kinraide and Behan (1975) noted that as sections decreased in size, O_2 uptake increased, and organic acid accumulation was depressed. The specific inhibitors of the citric acid cycle, malonate and fluoroacetate, inhibited the increased oxygen consumption and restored acid accumulation in the thin, 1-mm leaf sections. Other respiratory inhibitors, not specific for the citric acid cycle, did not restore acid accumulation. Hence, they concluded that respiration competes with acid synthesis.

Szarek measured increased oxygen consumption in the light, or perhaps a decreased oxygen evolution, with increasing acid content in *Opuntia* (Szarek and Ting, 1974b).

Osmond and Björkman (1975) measured both dark and light fixation of CO_2 in *Kalanchoe daigremontiana* as a function of O_2 concentration. Consistent with earlier reports, there was no inhibition of dark CO_2 fixation with increasing O_2 tension; however, O_2 inhibits light CO_2 fixation. This is further discussed in the Chapter 5.1.4.4.

We can conclude from the published data that O_2 is necessary for dark acidification, but high concentrations of O_2 do not significantly inhibit. Light deacidification may be more closely tied to oxygen tension; oxygen inhibits malate consumption. Osmond and Björkman suggest this may be related to the oxygen inhibition of photosynthesis. Apparently, malate decarboxylation during deacidification is tightly coupled to photosynthetic carboxylation. The latter is inhibited by oxygen, and pari passu, so is the former.

4.3.5.2 Carbon Dioxide

CO_2-Free Air. Effects of external CO_2 concentrations on acidification and deacidification during CAM have been discussed in detail by Wolf (1960). It is now well established that CO_2-free air inhibits nocturnal malic acid synthesis and accumulation (Kluge, 1968b, 1969b; see also Wolf, 1960). It is reasonable to assume that this effect is due to the lack of substrate for CO_2 dark fixation. However, with CO_2-free air malic acid accumulation at night is not prevented completely (see for example Kluge, 1969b) because of endogenous (respiratory) CO_2, which may serve as the substrate of dark CO_2 fixation.

There are indications that deacidification in the light is slightly stimulated in CO_2-free air (Wolf, 1960).

Atmosphere Enriched with CO_2. Bonner and Bonner (1948) and Thomas and Ranson (1954) found that at night enrichment of the atmosphere with CO_2 increased the rate of synthesis and the final level of malic acid. A CO_2 concentration of 2% was optimal. There was no enhancement of nocturnal malic acid synthesis with respect to normal air (0.03% CO_2) in 10–15% CO_2. CO_2 concentrations higher than 20% were clearly inhibitory if compared with normal air. Finally, at 70%, nocturnal malic acid accumulation is completely prevented.

Wood (1952) showed the enhancing effect of CO_2 concentration up to 10% on dark malic acid synthesis to be temperature-dependent. There was greater stimulation at 6° C compared with 19° C.

The above CO_2 effects might be interpreted as follows: Slight enrichment of the atmosphere increases the supply of external substrate for dark CO_2 fixation. Higher CO_2 concentrations, however, might cause stomatal closure (see Raschke, 1975a) excluding external CO_2 from the sites of β-carboxylation, or the CO_2 might act as a direct inhibitor of PEP carboxylase (see Chap. 4.2.1.1).

From experiments of Bruinsma (1958) and others (see Wolf, 1960) it is well known that 30% CO_2 prevents deacidification in the light. Nishida (1977) found inhibition of deacidification in 5% CO_2. Wolf (1960) and Beevers et al. (1966) interpret this as an indication that β-carboxylation can proceed in the light as well as dark, provided that there is sufficient CO_2 to saturate the demand of the reductive pentose phosphate cycle for this substrate (see also Chap. 4.2.1.3).

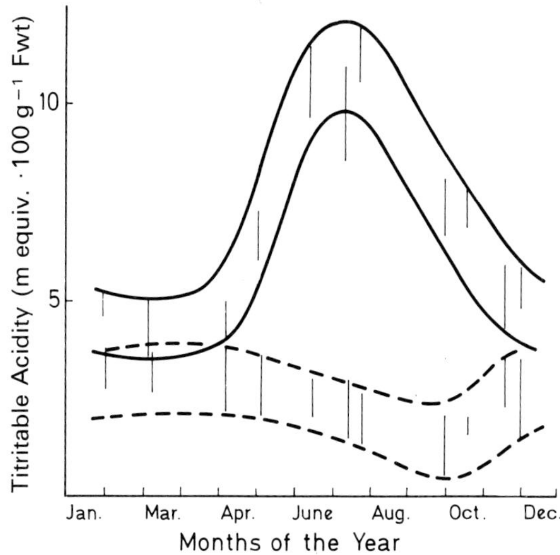

Fig. 4.9. *Sedum praealtum:* Maximum Δ of titratable acidity between evening (– – –) and morning (——) in response to the seasons (after Bennet-Clark, in Wolf, 1960)

4.4 Seasonal Control of CAM

It is well known that both the diurnal malic acid rhythm and the CO_2 exchange pattern of CAM undergo alterations during the course of the seasons (see Wolf, 1960, and Chapts. 5 and 6). Generally, maximum amplitude of the malic acid rhythm is found during the summer months (Fig. 4.9). This effect is probably due to the fact that during the summer CAM plants deacidify during the day to a lower level and reach a higher malic acid level during the dark period than during the rest of the year.

Obviously the control of CAM by the seasons is a multifactoral event. Photoperiod, thermoperiod, and hydroperiod are the most likely candidates which may affect and alter the CAM performance during the course of a year. Osmond (1978) suspects factors such as photoperiod or hydroperiod of accelerating normal developmental processes of CAM plants. Thus, effects of the above seasonal factors might act on CAM indirectly via development (see Chap. 4.4).

In the following, the influences of some seasonal factors on CAM will be considered in more detail.

4.4.1 Photoperiod

It has been known since the early 1940's that the CO_2 metabolism of certain CAM plants, viz., the short-day plant *Kalanchoe blossfeldiana* responds dramatically to photoperiod (Bode, 1942). Studies by Schmitz (1951) and Gregory et al. (1954) showed that CAM activity (as measured by dark CO_2 fixation) was induced by short days and prevented by long days or light interruption of long nights (see Fig. 5.5). These and other studies (Wilkins, 1962a, b; Queiroz, 1974) have clearly shown that this photoperiod response is phytochrome mediated.

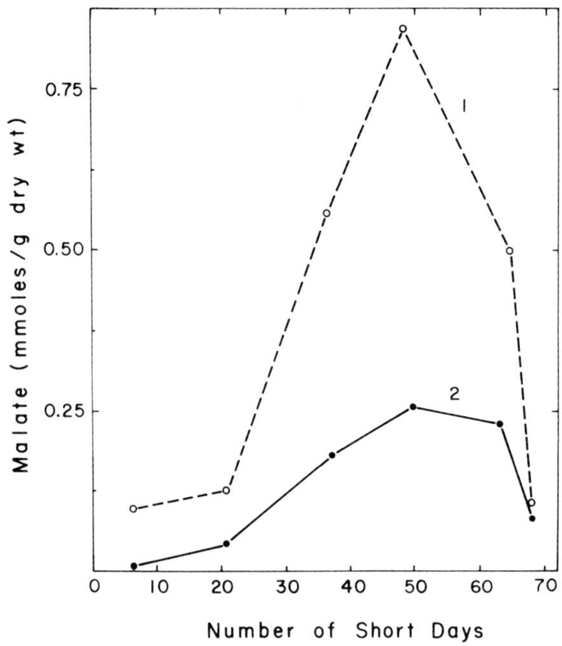

Fig. 4.10. Effect of short days on the variation in malic acid content in *Kalanchoe blossfeldiana* at beginning of the night (2) and at the end of the night (1) (after Queiroz and Morel, 1974, by permission)

Queiroz working at Le Phytotron (Gif-sur-Yvette) has extensively studied the photoperiodic response of *Kalanchoe blossfeldiana* (see the review of Queiroz, 1974). In general, the group in Gif-sur-Yvette found that in young leaves of *K. blossfeldiana* var. Tom Thumb, the CAM pathway was not operative under long-day conditions. There was no detectable dark CO_2 fixation, and only a small pool of malic acid. There was little detectable activity of PEP carboxylase, and malate enzyme and aspartate amino transferase were low when extracted and assayed in a fresh homogenate. However, if treated with polyethylene glycol, which presumably removes tannins and tannin-like substances, activity could be restored. They concluded, therefore, that the CAM enzymes were present, but inactive in situ.

When these long-day plants are transferred to short days and after an induction period of about 7 days, dark CO_2 fixation commences. Concomitantly there is a marked diurnal fluctuation of malic acid (Fig. 4.10) and extractable enzyme activity increases. Furthermore, there is observed a circadian oscillation of CO_2 exchange, malic acid content, and enzyme capacity (i.e., estimated maximum velocity).

PEP carboxylase induction shows a 7-day lag period, then an increase up to about 60 short days followed by a sudden decrease. Malate enzyme increases for about 20 days, then stabilizes at a level lower than PEP carboxylase. Aspartate aminotransferase increases continuously on short days, reaching an activity higher than PEP carboxylase, but lower than malate dehydrogenase. Oscillations of PEP carboxylase and aspartate aminotransferase are in phase but out of phase with malate enzyme, consistent with the hypothesis that PEP carboxylase synthesizes malate and malate enzyme degrades it (Fig. 4.6).

Because of the restoration of activity by polyethylene glycol as well as for other reasons, Queiroz's group has postulated a regulatory inhibitor of CAM. In vitro studies suggest that this inhibitor is dialyzable and thermostable. It inhibits, in vitro, PEP carboxylase, malate dehydrogenase, and aspartate aminotransferase, but not malate enzyme (Morel et al., 1972).

Although these studies are quite complex and not completely understood, the results demonstrate a real photoperiodic response of CAM metabolism reflected in CO_2 exchange, enzymic activity, and malate accumulation.

Osmond et al. (1973) showed a ^{13}C isotope composition shift from $-26.4‰$ to -18.4 to $-19.1‰$ when *K. blossfeldiana* was shifted from 15-h days to 9-h days.

It should be mentioned that no CAM species other than *Kalanchoe blossfeldiana* is known in which CAM is so drastically affected by the photoperiod. Thus, *K. blossfeldiana* might be considered as a special case, although some photoperiodic effects on gas exchange have been described in *K. daigremontiana* (Marcelle, 1975; see Chaps. 5.1.3 and 5.1.4). Crews et al. (1976) reported that *Aloe vera* functioned as a CAM plant most effectively on 9-h days, however, not in continuous light. In this plant, concomitant with organic acid fluctuation, PEP carboxykinase increased in activity. Recent results of Chaturvedi and Zabka (1977a, b) suggest interactions between photoperiodic induction of CAM in *Kalanchoë blossfeldiana* and water stress.

4.4.2 Thermoperiod

It is generally accepted that low night and high day temperatures are correlated with enhanced CAM activity (e.g., Vickery, 1954; Bennet-Clark, 1933a, b; Queiroz, 1965, 1966).

Queiroz (1965) using *K. blossfeldiana* plants grown in the Gif Phytotron showed that when transferred from short days to long days, the optimum day/night temperatures for acid accumulation were 27°/17° C. Night temperatures above and below 17° C were associated with lower acid accumulation. In addition, Queiroz (1966) showed that short-day plants grown at light temperatures of 27° C accumulated more acid than those grown at 17° or 12° C. Hence, he has concluded that for these plants, the optimum CAM activity is at 27°/17° C day/night cycles. $\delta^{13}C$ shifts upward in these plants, consistent with the shift from C_3-like to CAM (Allaway et al., 1974b).

Neales (1973b), using pineapple (*Ananas comosus*) demonstrated a correlation between the ratio of dark to light CO_2 fixation and night temperature. As night temperature increased (day at 27.5° C) from 15° to 35° C, there was less night-time CO_2 fixation relative to day-time fixation (see Chap. 5.1.3).

Crews et al. (1976, see also Joshi and Bartakke, 1974) reported that CAM was most effective in *Aloe vera* when grown under warm days (30° C) and cool nights (15° C) and short photoperiods (9-h days). Constant temperatures of 15° or 30° C and continuous light eliminated organic acid fluctuation. Of several enzymes studied, only PEP carboxykinase activity correlated with acid fluctuation.

Research of Szarek et al. (1973) showed that in *Opuntia basilaris* with closed stomata and no exogenous gas exchange, high temperatures resulted in higher dark malic acid levels. For 12 h in the dark at 32° C, acid was 34.2 ± 4.1 μeg/g while at 15° C it was 16.4 ± 4.5 μeg/g. Although these data appear to contrast with other reports, a near-linear increase in CO_2 produced by respiration was measured in *O. basilaris*, and hence the increased partial pressure of CO_2 trapped within the tissue undoubtedly accounts for the greater malic acid accumulation. Contrary to this, Bennet-Clark (1933a, b) found a decreased malic acid accumulation in *Bryophyllum* with an increase in respiration, but we assume here that stomata were not limiting exogenous CO_2 exchange.

As indicated earlier, Brandon (1967) reported that the temperature responses of PEP carboxylase and malate enzyme were such that at low temperatures acid could tend to accumulate, and Queiroz (1968a) showed that the time of day at which PEP carboxylase and malate enzyme were maximum and minimum was a function of the thermoperiod. Hence, the thermal properties of the known CAM enzymes may account in part for the known thermoperiodic responses.

An additional complicating factor is the stomatal response of CAM plants to temperature. Work at Riverside demonstrated that with *Kalanchoe blossfeldiana*, when grown under cool thermoperiods or with little diurnal fluctuations of temperature, stomatal opening and water loss were greater during the day than the night. With cool nights and warm days (30°/15° C) the CAM reverse-phase stomatal opening appeared (Fig. 7.8). Shreve (1916) also reported high temperature night-time stomatal opening in *Opuntia vesicolor*.

4.4.3 Hydroperiod

The close dependence of water status on temperature makes a separate consideration of hydroperiod independent of thermoperiod difficult. High temperatures tend to cause water stress and concomitant stomatal closure. Stomatal closure, if both day and night, will restrict metabolism.

As discussed earlier in detail (Chap. 4.3.4.2), there is now much information available confirming that in certain plant species water stress will cause a shift from non-CAM (i.e., C_3-like) carbon metabolism to CAM. On the other hand, there is no doubt that the majority of CAM plant species feature continually the CAM type of photosynthesis.

For example, the work of Ting's group at the Boyd Desert Research Center in Southern California has shown that cacti with water potentials of −12 to −15 bars perform CAM when growing under very arid desert conditions (Szarek et al., 1973; see Chap. 6.2.2.1). When water potentials increase in the range of −5 bars, stomata open at night, exogenous CO_2 is taken in, and there is an increase in diurnal organic acid fluctuation. Under the usual arid conditions of Southern California deserts, little or no daytime CO_2 fixation occurs. These data show the dependence of water status and hydroperiod on CAM activity, and suggest that the control here is largely stomatal.

Further effects of hydroperiod, in particular on CAM performance at natural stands, are discussed in Chapter 6.

4.5 Developmental Control of CAM

It is well known that the capacity for CAM increases with leaf age (Wolf, 1960; Avadhani et al., 1971; Jones, 1975; Osmond, 1978). In *Mesembryanthemum crystallinum*, induced CAM appears first in mature leaves and only later in younger leaves as they develop (Winter and Lüttge, 1976; von Willert et al., 1976a; Mooney et al., 1974; and Chap. 6.2.2.3). As indicated earlier in this chapter, Osmond (1978) has outlined that induction of CAM by low water potentials or by photoperiod may reflect an acceleration of normal developmental processes by the above factors. From this point of view it can be argued that a distinction between constitutive and facultative CAM plants (see Chap. 3.4) might be without relevance.

Furthermore, it should be mentioned that many of the results obtained by Queiroz's group on CAM control by photoperiods in *K. blossfeldiana* are based on investigations of the upper, young leaves and not the mature ones which represent the majority on the plant. Thus, it must be asked if the very interesting induction phenomenon is relevant to CAM of the whole adult *K. blossfeldiana* plant.

4.6 Conclusions

The available data strongly indicate a dependence of CAM activity on environmental signals (see also Chaps. 5 and 6). Photoperiod, thermoperiod, and hydroperiod all influence and control CAM in a complex manner. It seems reasonable to conclude that these environmental signals can reversibly shift certain CAM plants from the CAM type to the C_3 type of photosynthesis, and also can regulate the degree of CAM activity when featuring CAM.

Libbert (1976) has defined the terms "control" and "regulation" rather precisely. According to his definition, control refers to the effect of information on the flow (of energy, mass, etc.) through an open system in which the output of the system does not influence the flow. Regulation, however, in the strict sense is a closed control system where the effect of the control is measured at the output of the system and where the measured value affects the flow with a feedback loop (see Fig. 4.11). Such regulation tends to maintain flow of energy or mass at a constant level. Because it is a characteristic of CAM that steady state is rarely reached, we tend to believe that the regularity of CAM is maintained more by control processes rather than regulatory processes. This does not necessarily imply, however, the exclusion of partial metabolic sequences of CAM (for example: feedback regulation of PEP carboxylase by malate; regulation of glycolysis), or distinct physiological systems such as stomata which may in fact represent regulatory circuits in the above sense (see Raschke, 1966).

To a large extent the shifts in metabolic options in CAM plants are a direct function of water status and it is this, of course, which seems to be the adaptive significance of CAM. Nevertheless, it seems clear, particularly in the case of the photoperiodic studies, that factors other than water status are important.

These responses to photo-, thermo-, and hydroperiods tend to give a distinct seasonal asepct to CAM. Although no generalizations can be made with respect to seasonal patterns, there are quite definite species-specific patterns. The best

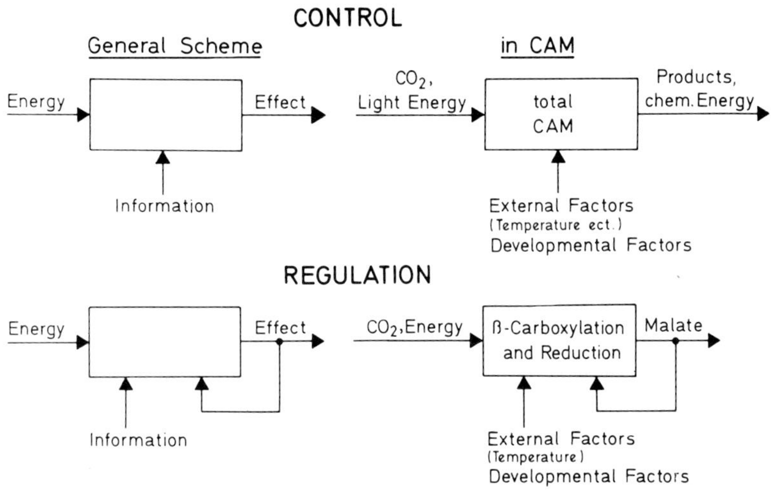

Fig. 4.11. General diagrams (*left*) characterizing control and regulation (after Libbert, 1976), and the application of these diagrams on CAM (*right*)

example would be the photoperiodic response of *Kalanchoe blossfeldiana* var. Tom Thumb.

There would also be a seasonal response to thermo- and hydroperiod insofar as there were predictable occurrences in a particular environment.

Finally, it should be made clear that the Cactaceae type of CAM shift is quite different from the shift reported in the Aizoaceae, Portulacaceae, and certain Crassulaceae. In the latter, the response to photoperiod and water stress appear to be a real shift in metabolic pathway. Not only do acids begin to fluctuate concomitant with gas exchange at night, but enzymes change quantitatively. Hence, the evidence suggests a reversible shift from C_3 to CAM.

In the case of the Cactaceae and many of the Crassulaceae, there is no evidence for a C_3 photosynthesis to CAM shift, but these plants move from an idling phase in which stomata are closed and a slight CAM metabolism persists to CAM where water is available. Excess water through irrigation did not result in C_3 photosynthesis although some daylight stomatal opening is observed. The latter could contribute to some additional ^{13}C fractionation by RudP carboxylase and slightly lower $\delta^{13}C$ estimates.

In any case, CAM plants such as cacti in their natural environments respond to water by exogenous nocturnal CO_2 exchange and active CAM metabolism. Many other succulent species respond to water stress by shifting from C_3 photosynthesis to CAM, thus conserving water.

Hence, in response to water, succulents show the full range from hermetically sealed, recycling CO_2 through the CAM pathway, to "full" CAM with nocturnal CO_2 uptake, and exclusive diurnal CO_2 assimilation through the C_3-pathway. Furthermore, it is not uncommon for some, e.g., the Crassulaceae, to show both nocturnal and diurnal CO_2 uptake and water loss.

Note: For some important recent papers dealing with the subjects of Chap. 4, in particular with endogenous rhythms of enzyme activities, properties of PEP carboxylase and intracellular malase transport, the Appendix on pp. 199—201 should be consulted.

5. Gas Exchange of CAM Plants

5.1 CO_2 Exchange

5.1.1 History

Historical aspects of gas exchange in CAM plants have been reviewed by several authors (Wolf, 1960; Ranson and Thomas, 1960; Beevers et al., 1966; Neales, 1975; Osmond, 1978). The net CO_2 uptake occurring at night is the essential and most outstanding feature of gas exchange in CAM plants. As early as 1804, De Saussure observed that stem sections obtained from an *Opuntia* species were capable of removing both oxygen and carbon dioxide from a closed atmosphere in the dark. The key sentences of De Saussure's publication (1804) are cited here in detail (cited from Wolf, 1960) because of their fundamental importance[2]: "Le carbonique en même raison que le gaz oxygène, lorsque le premier y est mêlé en petite quantité.

Les expériences que j'ai rapportées en A, sur l'inspiration du gaz oxygène, ont été faites dans un air que avait été preliminairement dépouillé de gaz carbonique, et l'on pourrait demander si le dernier n'est point absorbé par les feuilles dans l'air atmosphérique naturel préférablement aux gaz oxygène. J'ai en conséquence placé un cactus sous en récipient plein d'air commun, auqel j'avais ajouté du gaz carbonique, et j'ai trouvé que ce dernier et le gaz oxygène avaient été absorbés en même raison. L'inspiration seulement a été plus prompte que s'il n'y eût pas eu du gaz acide. Voici le détail d'un de ces expériences: 153.6 centimètres cubes ($7^3/_4$ pouces cubes) de cactus, ont été placés après midi dans $43^1/_2$ pouces cubes d'air confiné par du mercure; cet air, que je suppose égal à 100 parties, contenait 74 parties de gaz azote, 19 de gaz oxygène, et 7 de gaz acid carbonique. Après

[2] The cactus takes up CO_2 and oxygen in the same proportions when the former is present in small amounts.
Experiments reported in A. on oxygen fixation have been performed in air previously deprived of CO_2, and the question arises whether this latter is taken up by leaves in the normal air preferably to oxygen. I could show that a cactus placed in a jar with air enriched in CO_2 fixes CO_2 and oxygen at similar rates. Fixation was faster than in the absence of added CO_2. The experiment was as follows: 153.6 cm^3 ($7^3/_4$ cubic inches) of cactus were placed in the afternoon in a jar containing $43^1/_2$ cubic inches of air over mercury; the air contained 74% nitrogen, 19% oxygen and 7% CO_2 in volume. After 12 h darkness, the volume of the air was reduced by the plant to 81% of the initial volume. Hence, air uptake by the leaves was about $1^1/_4$ times of their volume. I found by analyzing the remaining air that the leaves absorbed $13^1/_2$ parts of oxygen and $5^1/_2$ part of CO_2. The two gasses were thereafter fixed in similar proportions, about $^3/_4$ of their initial amount. After their fixation the leaves were saturated and no further fixation could be observed in continuing darkness. (Translation from French by Dr. Orlando Queiroz.)

douze heures de séjour à l'obscurité, le volume d'air était reduit à 81 parties. Les feuilles avaient donc fait une inspiration égale à environ une fois et $^1/_4$ leur volume. J'ai trouvé par l'analyse de l'air restant qu'elles avaient inspiré $13^1/_2$ parties de gaz oxygène et $5^1/_2$ parties de gaz acide. Elles avaient donc inspiré les deux gaz en même raison, ou les $^3/_4$ de l'un et de l'autre. Après cette inspiration elles étaient saturées; elles n'ont pû continuer à diminuer leur atmosphère par une obscurité prolongée."

Thomas (1949) and Wolf (1949) named this anomalous CO_2 fixation occurring during the dark period, in honor of its discoverer, the De Saussure effect. The term De Saussure effect should not be used synonymously with CAM, but be reserved to characterize the phenomenon of dark CO_2 fixation, or, more precisely, the phenomenon of net CO_2 uptake from the atmosphere during the night by plants.

During the early 19th century, the discovery that certain plants may fix CO_2 at night was unexpected and very difficult to fit into the developing concept of how plants gain carbon. Hence, neither De Saussure nor other plant physiologists of his own or following generations were able to interpret the effect correctly or appreciate its full importance. Because of these circumstances, De Saussure's discovery was forgotten and unnoticed for more than a century. As a consequence, the key was lost which would have solved the riddle of how certain succulent plants accumulate malic acid during the night, a phenomenon which was discovered by Heyne (see Wolf, 1960), only ten years later than De Saussure's first observation of dark CO_2 fixation by plants (see Chap. 3).

It was not before 1949 that Thomas (1949), and Wolf (1949), stimulated by the discovery of nonphotosynthetic CO_2 fixation in propionic bacteria by Wood and Werkmann (1938), recognized dark CO_2 fixation as a missing link in the causal chain leading to dark malic acid accumulation during CAM.

Further evidence for the concept that dark CO_2 fixation is involved in CAM was provided by Thurlow and Bonner (1948) who demonstrated the appearance of radiocarbon in malic acid when $^{14}CO_2$ was supplied in the dark (see Chap. 3).

New insights were obtained when Infrared Gas Analyzers (IRGA) became generally available. These instruments allowed for important methodological progress in studying CO_2 exchange of CAM plants. The IRGA technique, with an open system, measured net CO_2 uptake and production continuously during complete day/night cycles or even longer periods of time, under different experimental conditions. Thus, with the use of the IRGA technique, ecologically relevant quantitative data on the carbon balance of CAM plants became available.

As far as is known, Gregory et al. (1954) and Nuernbergk (1955, 1957) were the pioneers in applying the techniques of infrared gas analysis to the problem of CAM-linked CO_2 exchange. Based on the data of such measurements, Nuernbergk (1961) stated that it is the net fixation of atmospheric CO_2 occurring during the night which provides most of the carbon gain in CAM plants.

5.1.2 General Phenomena of CO_2 Exchange

A "standard diurnal pattern" of CO_2 exchange during an experimental dark/light regime of 12:12 h and under constant temperatures of 20° C can be described. Such curves have been extensively discussed by Nuernbergk (1961, 1962),

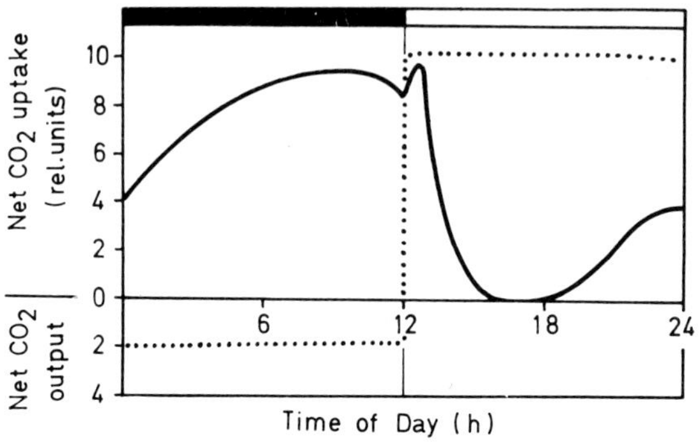

Fig. 5.1. Schematic standard curves of CO_2 exchange in a CAM plant (——) and a C_3 plant (······)

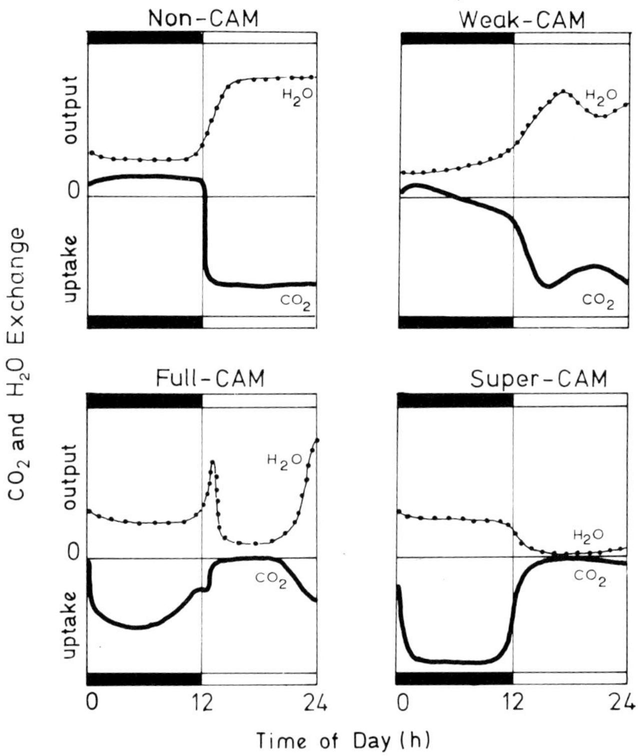

Fig. 5.2. Schemes of gas-exchange patterns as performed by CAM plants (after Neales, 1975, by permission)

Kluge and Fischer (1967), Kluge (1968a, b), Neales (1975) and others. A standard curve of CAM-linked CO_2 exchange is shown in Figure 5.1.

a) During the night period, net fixation of atmospheric CO_2 occurs which may equal or exceed the uptake of CO_2 during the day. This dark CO_2 fixation usually increases from a relatively low level at the beginning of the night, proceeds through a maximum, and decreases towards the end of the dark period.

b) During the day, the CO_2 exchange of CAM plants is characterized by three main events: (1) an initial increase in CO_2 uptake during the first one or two hours after the onset of light (CO_2 burst); (2) a decrease in net CO_2 fixation towards the compensation point where the curve may remain for periods of some hours; and (3) a final increase in CO_2 fixation at the end of the day.

As will be shown later in more detail, under the influence of a variety of exogenous and endogenous factors this standard pattern of CAM-linked CO_2 exchange may undergo variations. In this regard, Neales (1975) defined three types of CAM gas exchange (Fig. 5.2) ranging from the type shown in Figure 5.1 to a type which shows minimal nocturnal CO_2 fixation which then occurs at the end of the dark period. However, because any one CAM plant may show variations of the gas exchange pattern as a factor of environment, and because the different patterns above are not quantitatively defined, in this book we shall avoid clarification at this point. Suffice it to say that a variety of relative patterns are possible.

In the following sections, the gas exchange involved in CAM and the factors affecting it will be considered in more detail.

5.1.3 Patterns of CO_2 Exchange in the Dark

5.1.3.1 General Characteristics

As shown in Chapter 3 (see also Fig. 5.1), in most cases dark CO_2 fixation in CAM means net fixation of atmospheric carbon dioxide. However, this statement does not exclude that endogenously produced CO_2 (respiratory CO_2) may also serve as the substrate for dark CO_2 fixation. In special cases, mainly under the influence of environmental factors, endogenously produced CO_2 may provide the only source for dark CO_2 fixation and malic acid synthesis (cf. Chap. 6.2.2.1).

As can be seen from Figure 5.1, and as discussed above, the course of net dark CO_2 fixation reaches a maximum following a generally smooth curve. There is considerable evidence (see Chap. 5.3.2.3) that this typical shape of CO_2 exchange during the night is only partially controlled directly by the behavior of stomata. Major controlling factors in gas exchange at night are localized in the mesophyll cells of the chloroplast-containing tissue (Kluge and Fischer, 1967). The decrease of net CO_2 exchange occurring towards the end of the dark period can be explained in terms of increasingly exhausted storage capacity in the mesophyll cell vacuoles for the end product of CO_2 fixation, i.e., malic acid. As discussed in Chapter 4.3.2.3, the cytoplasmic malic acid pool is expected to increase when the storage vacuoles are no longer able to act as a sink for malic acid because the vacuolar malic acid concentration has reached its maximal value. An increase of the cytoplasmic malic acid concentration, however, would finally inhibit PEP carboxylase (see Chap. 4.3.2), which causes a decrease in the rate of CO_2 uptake from the atmosphere. This argumentation is supported by the fact that malic acid accumulation proceeds more or less linearly only during the first 10 h or so of a reasonably extended dark period. After this time, the malic acid accumulation becomes saturated (see Wolf, 1960). This phase is correlated in time with the decreasing dark CO_2 fixation rate (cf. Thomas and Beevers, 1949).

5.1.3.2 Factors Affecting CO_2 Exchange During the Dark Period

a) Nocturnal Temperature

There is overwhelming evidence that high night temperatures inhibit malic acid accumulation in CAM plants, whereas low night temperatures have the opposite effect (see reviews of Wolf, 1960; Ranson and Thomas, 1960; see Chap. 4.3.1). Because malic acid is the end product of the metabolic pathway of dark CO_2 fixation, it is reasonable to predict that CO_2 exchange during the night would also be affected by temperature.

Indeed, there are numerous observations documented from a variety of CAM plants that dark CO_2 fixation in CAM is generally lowered by high night temperatures (Nuernbergk, 1962; Kluge, 1968; Neales, 1973a, b; Kluge et al., 1973; Patten and Dinger, 1969; Dinger and Patten, 1972, 1974; Despain et al., 1970; Troughton et al., 1974; Neales, 1975; Lange et al., 1975; Allaway et al., 1974b). Temperatures higher than 35° C eliminate the nocturnal net CO_2 fixation and shift the gas exchange into the region of net CO_2 production (see e.g., Nuernbergk, 1962; Kluge, 1968b; Kluge et al., 1973; Neales, 1973a; see Fig. 5.9). However, considerable differences seem to exist between certain species of CAM plants with respect to temperature sensitivity of dark CO_2 fixation. For example, in *Caralluma negevensis*, temperatures of about 22° C prevent net CO_2 fixation during the night (see also Chap. 6). This is true also in three species of *Echinocereus* (Dinger and Patten, 1974). In contrast, orchids growing in Singapore where the night temperatures never falls below 20° C, perform CAM well at this temperature (Neales and Hew, 1975; Avadhani, personal communication).

Net dark CO_2 fixation is also inhibited by very low night temperatures. In *Kalanchoe daigremontiana* and *K. tubiflora*, at about 5° C nocturnal net CO_2 uptake is zero (Nuernbergk, 1962; Kluge, 1969c). The same effect has been observed in *Tillandsia usneoides* (Kluge et al., 1973). In *Echinocereus*, Dinger and Patten (1974) reported that net dark CO_2 fixation at 5° C was only slightly inhibited.

In most of the CAM plants investigated, substantial net CO_2 fixation occurs if the night temperatures are 10° C or above. Optimal CO_2 uptake is observed at nightly temperatures ranging from 15°–22° C (see Fig. 5.3, Kluge et al., 1973; Medina et al., 1977; Neales, 1975; for exceptions see Dinger and Patten, 1974; Lange et al., 1975).

The interrelationship between night temperature and intensity of dark CO_2 fixation described above is summarized in Figure 5.3 which shows the results of gas exchange measurements under different temperature regimes in *Tillandsia usneoides*. There is little doubt that the results of these measurements may be generalized for the majority of CAM plants. For example, Neales (1975) has obtained nearly identical diagrams for *Agave americana* (see Fig. 5.9). Some of the exceptions deviating from this general behavior of dark CO_2 fixation under different high temperatures can reasonably be explained as thermal adaptations to special environmental conditions (cf. Patten and Dinger, 1969; Dinger and Patten, 1972, 1974).

The mechanism by which temperature affects dark CO_2 fixation in CAM plants cannot be entirely explained. Since net CO_2 uptake from the atmosphere

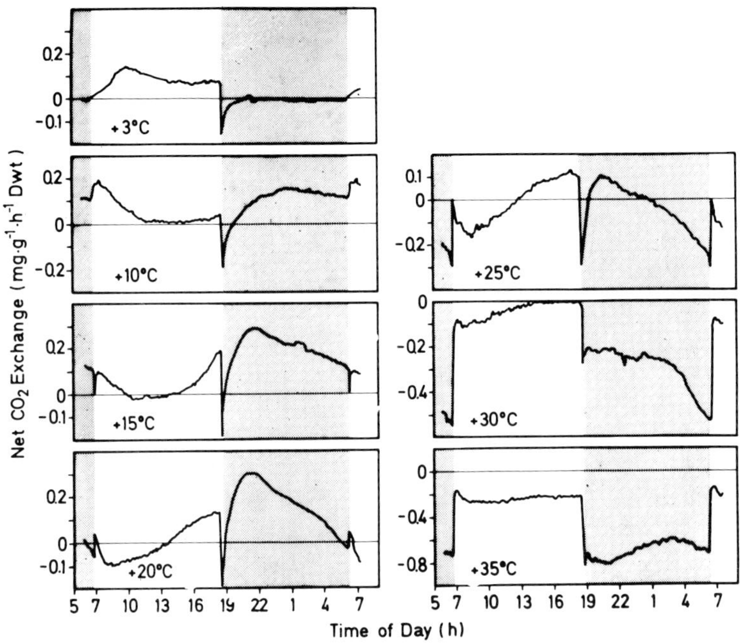

Fig. 5.3. Influence of temperature on CO_2 exchange (*Tillandsia usneoides*). Light intensity 30,000 lux (after Kluge et al., 1973)

during the night presumes nocturnal opening of stomata, it is suspected that stomata respond to temperature such that the fluxes of CO_2 would be influenced as described above. Indeed, this was the case for *K. blossfeldiana* stomata (Ting et al., 1967). However, it will be shown later in more detail, that in some cases stomata of CAM plants under normal conditions seem to be controlled by CO_2 exchange of the mesophyll tissue and not vice versa. A further argument against stomatal effects is the finding that high temperatures tend to open stomata (Schulze et al., 1974; Hall et al., 1976). Thus, an increase in net CO_2 uptake should result if the temperatures are high.

Brandon (1967) concluded from in vitro experiments that the thermal properties of the carboxylating and decarboxylating enzymes (PEP carboxylase, malic enzyme, see Chap. 4.2.3.2) are such that lower temperatures would favor CO_2 consumption (i.e., malic acid synthesis); higher temperatures, however, would increase CO_2 production by malic acid decarboxylation. Hence, the thermal properties of the enzymic proteins could account partly for the observed temperature effects.

Also, at high temperatures, CO_2 production by mitochondrial respiration increases drastically, thus shifting the gas exchange from net CO_2 fixation to net CO_2 production (Szarek and Ting, 1974b).

Kaplan et al. (1976a, b) using a highly stabilized oxygen electrode with a full-scale deflection of 400 µl per 1 O_2 against a background of 21% O_2 were able to resolve net dark CO_2 fixation from respiration. Assuming an $RQ = 1$, they calcu-

Table 5.1. Dark CO_2 fixation by *Kalanchoe daigremontiana* as a function of temperature (from Kaplan et al., 1976b)

Temperature °C	μg CO_2 g^{-1} dry wt.		
	Gross dark[a] CO_2 fixation	Respiration	Net dark CO_2 fixation
10	804	136	668
14	851	188	663
19	839	401	438
24	823	590	233

[a] Cumulative gross dark fixation, respiration, and calculated net dark CO_2 fixation integrated from onset of dark period to maximum rate of gross dark fixation.

lated respiration in terms of CO_2 production from O_2 consumption. In *K. daigremontiana*, as temperature increased from 10 to 24° C, the total cumulative dark fixation remained unchanged while dark respiration, i.e., CO_2 evolution, increased 3.6-fold. By calculation, net dark CO_2 fixation decreased 3.0-fold (Table 5.1). The observed increase in dark CO_2 fixation with decreasing temperature could be entirely accounted for by the temperature-dependent increased respiration.

The decrease of net CO_2 fixation observed at temperatures below 10° C could be explained by the lower enzyme activities which certainly can be expected at this temperature range.

Also the rate at which malic acid is removed from the cytoplasm by transport into the vacuoles should decrease at low temperatures because the flux of malic acid into the vacuole is assumed to be an active process (see Chap. 4.2.3.2). Hence, relatively high cytoplasmic malic acid concentration would result which should inhibit PEP carboxylase thus lowering CO_2 consumption.

b) Influence of Preillumination

Light Intensity. Early investigations of De Vries (1884) have clearly shown that nocturnal malic acid accumulation by succulents occurs only with sufficient preillumination during the period proceeding the night. From these results, it can be predicted that net dark CO_2 fixation will depend on the duration and intensity of the illumination during the preceding day (see Nuernbergk, 1957, 1961). However, there have been few attempts to verify this prediction. In those few cases where the influence of the intensity of preillumination on subsequent dark CO_2 fixation was investigated (Kluge et al., 1973; Kluge, unpublished; Nose et al. 1977), plants (*Tillandsia usneoides*, *Kalanchoe tubiflora*, *Ananas*) behaved as predicted (Fig. 5.4). After application of high light intensities at constant photoperiods, dark CO_2 fixation during the subsequent night was relatively high, and decreased if the intensity was diminished. It is most likely that other species of CAM plants behave similarly, but of course dose-response curves may differ among species.

Kaplan et al. (1976a, b) also reported net CO_2 fixation calculated from gross dark CO_2 fixation less dark respiration as a function of the preceding light intensity. When light increased from 3.2 to 12 mW cm^{-2}, dark fixation of CO_2 by

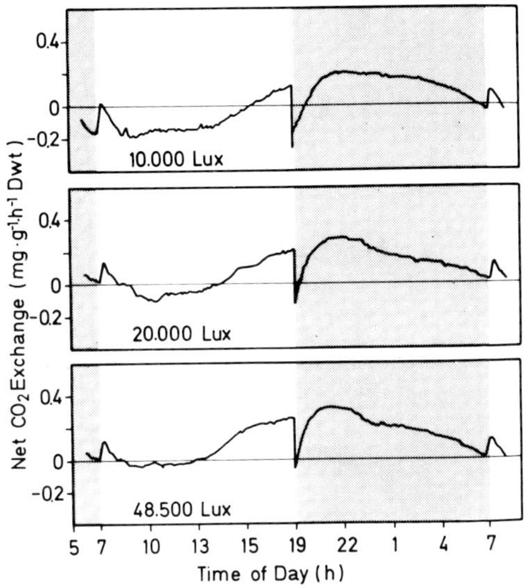

Fig. 5.4. Influence of light intensity on CO_2 exchange (*Tillandsia usneoides*). Temp. 20° C (after Kluge et al., 1973)

K. daigremontiana increased 2.5-fold, whereas dark respiration was not affected; hence, net dark CO_2 fixation increased 3.6-fold with a 3.8-fold increase in radiant energy supplied during its previous light period (Table 5.2).

There are several possibilities to explain the mechanism by which preillumination can influence subsequent dark CO_2 fixation. Nuernbergk (1961, 1962) postulated that during the day, NAD^+ is reduced and ADP is phosphorylated by light energy thus providing chemical energy in the form of NADH and ATP which could be used in dark CO_2 fixation during the following night. This postulate presumes that NADH and ATP are stored until the beginning of the dark period. A simple calculation may show that Nuernbergk's hypothesis cannot be valid: *Kalanchoe tubiflora* fixed in one night 0.2 mmol $CO_2 g^{-1}$ dry weight. To reduce CO_2 to the level of malate, the same amount of NADH would be required,

Table 5.2. Dark CO_2 fixation by *Kalanchoe daigremontiana* as a function of light intensity during the preceeding light period (from Kaplan et al., 1976b)

Light intensity $mW\ cm^{-2}$	µg CO_2 g^{-1} dry wt.		
	Gross dark[a] CO_2 fixation	Respiration	Net dark CO_2 fixation
12	1409	260	1149
8.1	1230	240	990
6	1013	238	775
3.2	564	245	319

[a] Cumulative gross dark fixation, respiration, and calculated net dark CO_2 fixation integrated from onset of dark period to maximum rate of gross dark fixation.

0.2 mmol. This estimate is more than four orders of magnitude higher than the maximal NADH concentration observed by Heber and Santarius (1965) in the chloroplasts and nongreen parts of spinach mesophyll cells.

In spite of these objections, there is no reason to reject Nuernbergk's basal idea that light energy is collected and stored for subsequent dark CO_2 fixation. However, instead of a direct accumulation of "assimilatory power" in the form of NADH and ATP, carbohydrates such as starch or related glucans (see Sutton, 1975a–c) are most likely involved in energy storage. During the following night, the stored glucan supplies the energy-rich CO_2 acceptor needed for dark CO_2 fixation (PEP) (cf. Chap. 3). Thus dark CO_2 fixation depends on photosynthesis, i.e., on preillumination. According to this hypothesis, light intensity would determine the amount of starch or glucan synthesized during the day, and thus the amount of PEP available for carboxylation during the night. This interpretation has recently been strongly supported by results of Kaplan et al. (1977).

A further possibility by which preillumination might act on dark CO_2 fixation is the regeneration of the storage capacity for malic acid. Substantial dark CO_2 fixation may occur only if "empty" vacuoles provide an effective sink for malic acid accumulation (see Chap. 4.2.3.2). Preillumination depletes the vacuoles of malic acid, thus enabling the cells to synthesize this substance via dark CO_2 fixation during the following night.

Photoperiod. There is strong evidence that photoperiod is of great significance in the regulation of nocturnal CO_2 exchange in certain CAM species. This influence of photoperiod can be demonstrated in the short-day plant, *Kalanchoe blossfeldiana* var. Tom Thumb (see Gregory et al., 1954; Spear and Thimann, 1954; Zabka et al., 1959; Nuernbergk, 1961; Queiroz, 1965). In this species, net dark CO_2 fixation occurs only after a critical number of short days (i.e., long nights). The magnitude of carbon gain during the night increases with the number of short days (see Fig. 5.5). Plants cultivated under long-day (i.e., short-night) conditions show little or no CO_2 uptake during the night.

If the long nights are interrupted by a light period of about 10 min, dark CO_2 fixation is inhibited. Thus, net dark CO_2 fixation in *Kalanchoe blossfeldiana* var. Tom Thumb is very similar to flower induction. Spear and Thimann (1954) postulated that dark CO_2 fixation was a precondition for flowering. However, Zabka and Edelman (1961 b), studying a different variety of *Kalanchoe blossfeldiana* (i.e., var. Feuerblüte), showed that dark CO_2 fixation was independent of photoperiodic flowering induction. In *Kalanchoe blossfeldiana* var. Feuerblüte, both short- and long-day plants fix CO_2 in the dark.

Also, in *Kalanchoe daigremontiana*, one of the classic species used in CAM research, both short- and long-day-treated plants have nocturnal CO_2 uptake and malic acid synthesis. However, the shape of the gas exchange curves differs between the two experimental treatments (Marcelle, 1975). Under short-day conditions net dark CO_2 fixation is maximal during the beginning of the dark period, and approximates the compensation point toward the end. In plants cultivated under long-day conditions, a higher net CO_2 fixation remains throughout the night (Fig. 5.6). The effects of photoperiod on nocturnal CO_2 exchange can be summarized as follows: in some cases, net dark CO_2 fixation is induced by short-

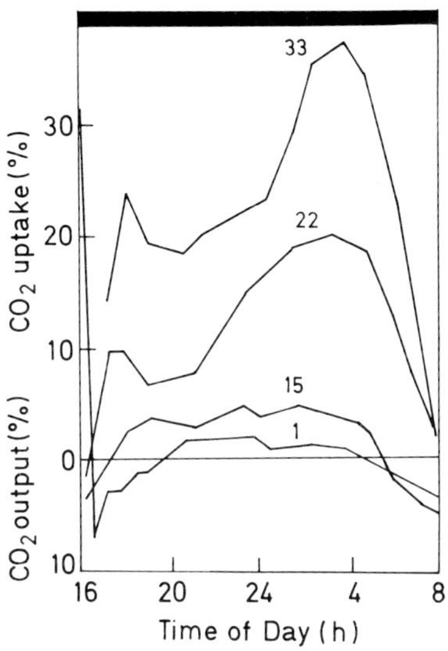

Fig. 5.5

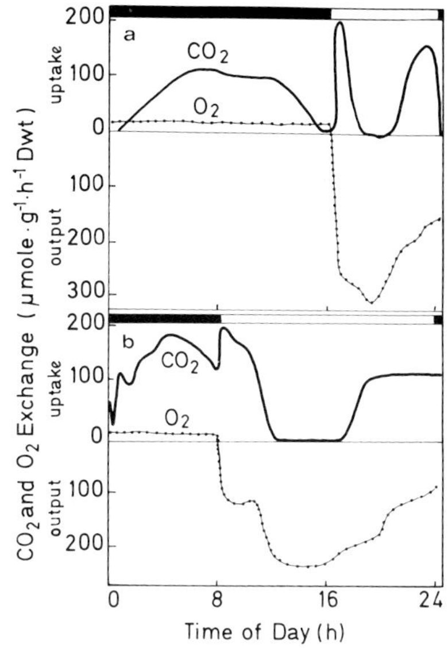

Fig. 5.6

Fig. 5.5. Dark CO_2 exchange in *Kalanchoe blossfeldiana*, precultivated under long-day conditions (16 l:8 d), as influenced by increasing numbers of short days (8 l:16 d) (modified after Gregory et al., 1954)

Fig. 5.6. CO_2 and O_2 exchange under long- and short-day conditions (*Kalanchoe daigremontiana*) CO_2——; O_2······ (after Marcelle, 1975, by permission)

day conditions (*K. blossfeldiana* var. Tom Thumb). In other cases, photoperiod has little influence (*K. blossfeldiana* var. Feuerblüte) on nocturnal gas exchange. The behavior of *K. daigremontiana*, where mainly the shape of the nocturnal CO_2 exchange curve is affected, can probably be regarded as representative for the majority of CAM plants. However, more investigations with a larger sample of species are needed prior to generalization. Interestingly, there are no observations of the long-day requirement for net dark CO_2 fixation by CAM plants. The metabolic aspects of CAM control by photoperiod have been discussed in Chapter 4.4.1.

c) *Influence of Drought*

Under natural or experimental drought periods, the uptake of atmospheric CO_2 by plants is usually reduced. This is true as well in dark CO_2 fixation by such CAM plants as *Kalanchoe daigremontiana*. In the latter, Kluge and Fischer (1967) showed decreasing carbon gain during the night if plants were treated by soil desiccation (see Fig. 6.1). Also, in *Echeveria* the amount of CO_2 fixed during the night was reduced under water stress (Meinzer and Rundel, 1973). In *Dudleya farinosa*, also Bartholomew (1973) observed a reduction in day net CO_2 uptake during soil desiccation. However, in the latter case contrastingly to many other CAM species, CO_2 uptake during the night was not reduced until the end of the drought period (see Chap. 6.2.2.1).

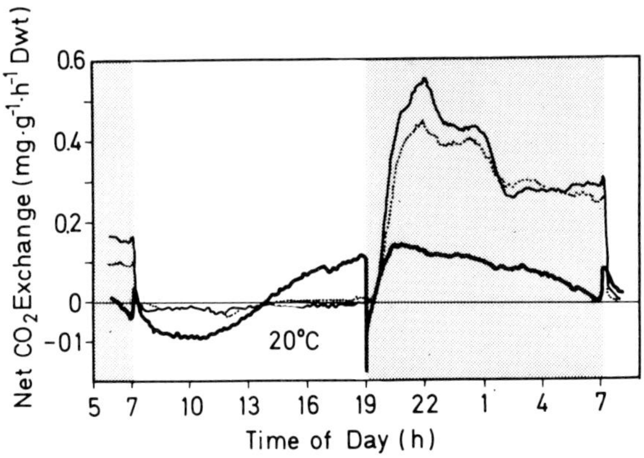

Fig. 5.7. Influence of water stress on CO_2 exchange of *Tillandsia usneoides* (after Kluge et al., 1973). —— plants with 80.2% water content of initial plant fresh weight; ······82.3%; —— 85.7%. Light intensity: 48,500 lux, temperature 20° C

Apparent stimulating effects of drought on dark CO_2 fixation have been observed in certain CAM plants. *Tillandsia usneoides* grown under drought conditions showed a higher CO_2 uptake during the night than well-irrigated controls (Kluge et al., 1973; Fig. 5.7). This behavior results in a higher carbon gain in water-stressed plants than in plants kept wet during cultivation (Table 5.3). Similar results were obtained by Benzing and Renfrow (1971), and Medina et al. (1977) in other Bromeliaceae.

Mesembryanthemum crystallinum and other Aizoaceae provide examples of net dark CO_2 fixation induced by water stress (for details see Chaps. 4.3.4.2 and 6.2.2.3). Under a normal water status, in these species CO_2 exchange is similar to a C_3 plant, i.e., without net dark CO_2 fixation. However, under drought (or salinity stress), the species develop all characteristics of CAM including dark CO_2 fixation (Fig. 5.8). Similarly if *Portulacaria afra* (Ting and Hanscom, 1977) is stressed by withholding water, there is a shift from diurnal gas exchange to nocturnal gas exchange (see Fig. 4.8). In arid desert conditions, unless after a period of precipitation, there is virtually no gas exchange by cactus (Szarek et al., 1973). When tissue water potentials are high (-5 to -2 bars), nocturnal CO_2

Table 5.3. Interrelationship between CO_2 balance and water content of the plant (*Tillandsia usneoides*)[a]

Water content of plant (% of fr. wt.)	CO_2 balance (mg $CO_2 \cdot g^{-1}$ dry wt. $\cdot d^{-1}$)		
	During night	During day[b]	Total over 24 h
80.2	+3.66	−0.15	+3.51
82.3	+3.44	−0.09	+3.35
85.7	+0.94	−0.08	+0.86

[a] Data from Kluge et al. (1973).
[b] Conditions during the experiment: Light: 48,500 lux, Temperature: 20° C.

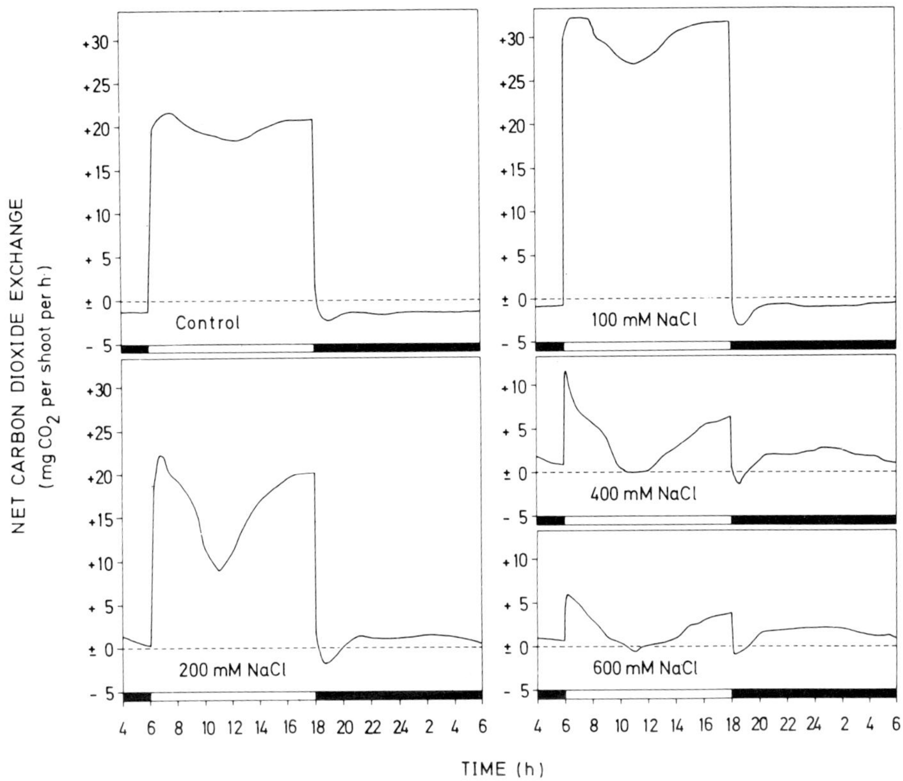

Fig. 5.8. *Mesembryanthemum crystallinum* after 14 days of irrigation with NaCl solutions of increasing concentrations. Positive values refer to net CO_2 uptake, negative values to net loss of CO_2 from the plant (from Winter and Lüttge 1976, by permission)

uptake occurs. When tissue potentials depress to −10 or −15 bars, nocturnal CO_2 uptake decreases (see Chap. 6).

Compared with dark CO_2 fixation, day CO_2 uptake by CAM plants seems more readily affected by water stress. With *Kalanchoe daigremontiana*, Kluge and Fischer (1967) observed a complete reduction of daytime CO_2 fixation after 4 days of drought, yet dark CO_2 fixation was still evident with only slight decrease even after 6 days of stress. Under drought conditions, many CAM plants take up CO_2 from the atmosphere exclusively at night (Fig. 6.1).

With extreme drought, cacti are capable of existing for extended periods with no exogenous CO_2 uptake by recycling endogenously produced CO_2 through the CAM pathway (Ting and Szarek, 1975, see also Fig. 6.7).

Attempting a causal analysis of drought effects on dark CO_2 fixation, it is reasonable to argue that the behavior of stomata might play a role. As in other plants, the stomata of CAM plants tend to close under water deficits, thus reducing the uptake of CO_2 (see Chap. 5.3).

Drought could also inhibit nocturnal CO_2 fixation by affecting the vacuolar storage capacity of malic acid. Malic acid is stored in the vacuoles as an aqueous solution, and water deficits might reduce the amount of malic acid which would be soluble in the vacuolar sap. Hence, drought may lower the storage capacity for malic acid.

Further explanations of drought effects on CAM are provided in Chapter 4.3.4. It should be noted, that irrigation converts certain CAM plants to daytime CO_2 uptake. This was shown for *Agave deserti*, where irrigation abolishes dark CO_2 fixation completely (Hartsock and Nobel, 1976).

5.1.4 CO_2 Exchange During the Light Period

5.1.4.1 General Characteristics

As previously outlined briefly, the standard curve of CO_2 exchange in CAM plants during the light period is characterized by three different phases: (1) the initial period of CO_2 uptake, (2) the depression of CO_2 uptake during the middle of the day, and (3) the final increase in CO_2 uptake.

Among these stages, only the depression of CO_2 uptake seems to be an essential feature of CAM. In contrast, as will be shown later in detail, both the initial CO_2 uptake and the final net CO_2 uptake may be missed in certain CAM plants.

There are numerous reports in the literature of a depression in daytime CO_2 uptake in non-CAM plants (see Stocker, 1960; Larcher, 1973). The occurrence of "noon depression of photosynthesis" (Stocker, 1960) seems to be especially common in plants growing in arid habitats, the same habitats in which CAM plants exist. However, there is a fundamental difference between the noon depression of CO_2 uptake in CAM and non-CAM plants. In species without CAM, the depression is caused by stress factors such as temporary water deficit, high air temperature, or reduced atmosphere moisture. These factors cause stomata closure during midday, thus inhibiting photosynthesis. In CAM plants, however, the occurrence of the CO_2 uptake depression is an essential consequence of malic acid metabolism which characterizes CAM (Kluge, 1968b). For example, in CAM plants the depression occurs independently of the actual water status of the leaves, i.e., it occurs in nonstressed plants as well as stressed. An explanation of the coupling between malate metabolism of CAM and the daytime closure of the stomata in CAM plants is suggested in Chapter 5.3.2.2.

5.1.4.2 Factors Affecting CO_2 Exchange During the Light Period

a) CO_2 Fixation and Malic Acid Accumulation of the Preceding Night

It is evident from the literature on CAM plants gas exchange that the magnitude of CO_2 fixation and malic acid accumulation during the previous night has a great influence on gas exchange during the following day (cf. Kluge, 1968b; Medina and Delgado, 1976). Generally, if the CO_2 fixation during the night has been high, it will be relatively low during the following day, and vice versa. Hence, all those factors which favor nocturnal CO_2 fixation will reduce the daytime CO_2 uptake from the atmosphere. In contrast, factors which inhibit nocturnal CO_2 uptake can be expected to have a positive influence on daytime CO_2 fixation. For

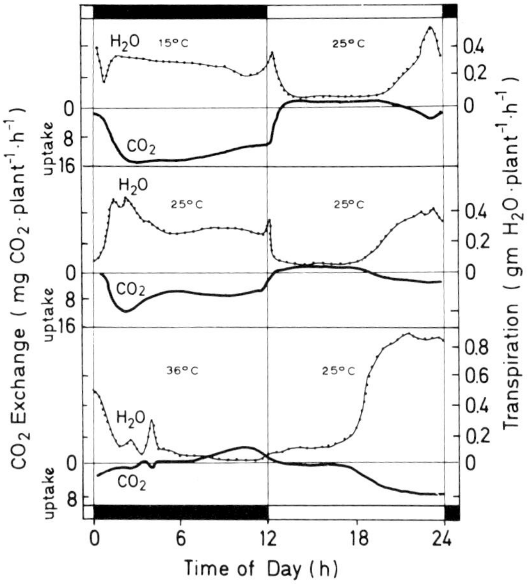

Fig. 5.9. The influence of night temperatures of 15°, 25° and 35° C on the pattern of CO_2 (——) and H_2O (······) exchange *Agave americana* (after Neales, 1975, by permission)

example, after relatively high or low night temperatures which prevent net nocturnal CO_2 fixation (see Chap. 5.1.3.2,a), CO_2 uptake during the following day is only weakly depressed or not affected (Fig. 5.9). Thus, CAM plants under certain conditions may have CO_2 exchange patterns nearly identical with C_3 plants (see Nuernbergk, 1962; Kluge, 1968b; Kluge et al., 1973; Neales, 1975; Ting and Hanscom, 1977). Nocturnal CO_2 fixation and concomitant malic acid synthesis is also inhibited by CO_2-free air during the night. After such treatment, the depression of CO_2 uptake during the following day is shorter than in controls (Kluge, 1968b; see Fig. 5.10).

Photoperiods unfavorable for dark CO_2 fixation or photoperiods in combination with unfavorable night temperatures will also reduce dark CO_2 fixation and malic acid synthesis, thus enhancing CO_2 uptake during the following day (Queiroz, 1965; Allaway et al., 1974b). CAM plants cultivated under such conditions show $\delta^{13}C$ values (see Chap. 3) tending in the direction of a C_3 pattern rather than of C_4 (Crews et al., 1976; Osmond et al., 1973). This reflects a high proportion of atmospheric CO_2 fixed directly by C_3-photosynthesis.

The question is how the intensity of nocturnal CO_2 uptake may influence CO_2 uptake during the following day. It is to be understood that the link connecting nocturnal and daytime CO_2 uptake is the malic acid accumulation during the night and disappearance of malic acid during the day. As outlined in Chapter 3.2, malic acid consumption in the light proceeds via decarboxylation to form a C_3-acid and CO_2. Thus, the accumulated malic acid provides an endogenous source of CO_2 for photosynthesis. It is reasonable to assume that CO_2 derived from malic acid could compete with atmospheric CO_2, since it is generated in the intimate vicinity of the photosynthetic apparatus. Hence, assuming that the pho-

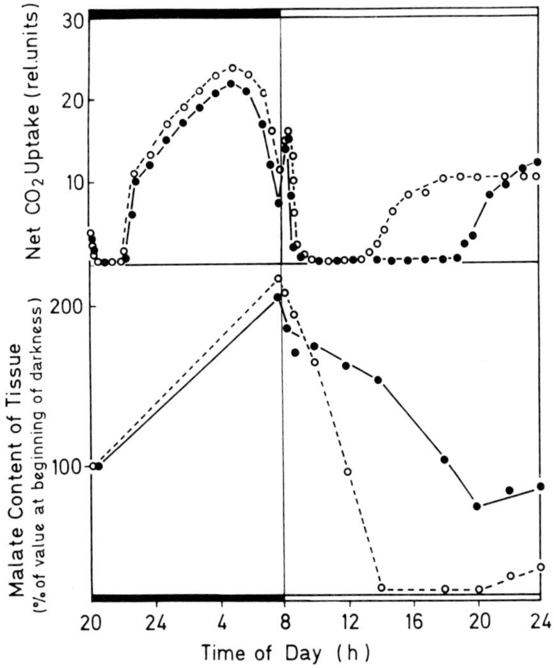

Fig. 5.10. Net CO_2 exchange and malate content in *Kalanchoe tubiflora*. Two individuals of a clone were held under identical conditions till the end of the dark period (8 h). During the light period given in the figure, the control plant was held at 8000 lux (●—●—●), the other one at 17,000 lux (○---○) (after Kluge, 1968b)

tosynthetic apparatus has a limited capacity, which is determined also by external factors such as light and temperature, it can be expected that the endogenously produced CO_2 will reduce the proportion of CO_2 which is fixed from the atmosphere. Thus, to some extent, the depression of CO_2 uptake during the day may be a consequence of the photosynthetic conversion of malic acid into carbohydrates with CO_2 as intermediate.

If this hypothesis is valid, the depression of CO_2 uptake should persist as long as endogenous CO_2 is produced during the day. In fact, there is convincing evidence that the length of the daytime CO_2 uptake depression depends mainly on the size of the malic acid pool (i.e., CO_2 pool) accumulated during the previous night, or from the velocity by which this pool is depleted (Kluge, 1968b). A typical example is given in Figure 5.10, which shows that there is a correlation in time between the depletion of malic acid and CO_2 uptake during the late phase of the light period. The mechanism of the correlation with stomatal opening is discussed in Chapter 5.3.2.2.

b) Light Intensity and Photoperiod

There are a few reports in the literature on the effect of different light intensities on the day CO_2 exchange of CAM plants. In *Kalanchoe daigremontiana* and *Ananas commosus*, high light intensity reduces the midday temporal extension of

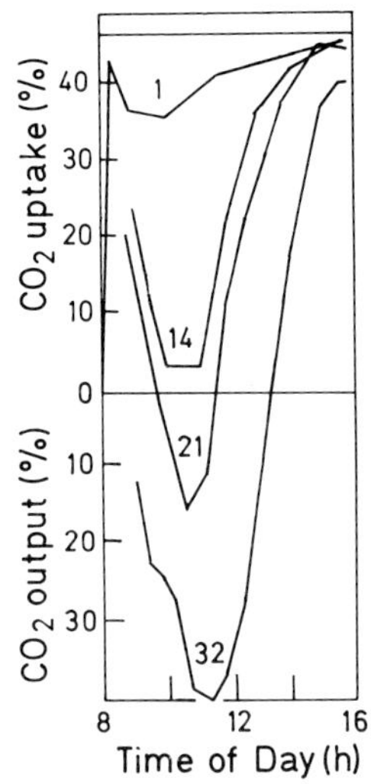

Fig. 5.11. Light CO_2 exchange in *Kalanchoe blossfeldiana*, precultivated under long-day conditions (16 l:8 d), as influenced by increasing numbers of short days (8 l:16 d) (modified after Gregory et al., 1954)

CO_2 uptake depression (Kluge, 1968b; see Fig. 5.10; Nose et al., 1977). On the basis of the hypothesis outlined above (see Chap. 5.1.4.2,a), this effect can be explained by assuming that high light enhances photosynthetic rates, thus increasing the photosynthetic consumption of malic acid. Under some experimental conditions with *Tillandsia usneoides*, there is net CO_2 output during the day (Kluge et al., 1973; see Fig. 5.4, see also Fig. 6.9 and 6.12). Carbon loss is greatest at low light intensity, and diminishes as the light intensity increases. However, even at 48,500 lux (the highest light intensity used in these experiments) net CO_2 output was still evident. In these experiments, therefore, it seems that only at high light intensities are photosynthetic rates great enough to prevent CO_2 loss from the tissue. The relatively high net CO_2 output shows that the decarboxylation of malic acid (i.e., CO_2 production) proceeds under low as well as high light. Thus, if light has a function in the mobilization of malic acid (see Wolf, 1960; Kluge, 1971b) it presumably requires lower light intensities than the photosynthetic refixation of CO_2. It is clear that considerations of light effects on gas exchange are necessarily complicated because of temperature effects.

CO_2 production from malic acid is also stimulated by high temperatures. Hence, the higher the day temperature, the higher the light that would be needed to fix the CO_2 produced from malic acid. It can be expected that CAM species from different habitats will differ significantly in this behavior.

Gregory et al. (1954) observed in *Kalanchoe blossfeldiana* that both dark CO_2 fixation and CO_2 uptake in the light period are affected by photoperiod. Plants precultivated under long-day conditions take up external CO_2 throughout the day with little depression. With increasing numbers of short days, a depression of daytime CO_2 fixation occurs, and finally even massive output of CO_2 may be observed (Fig. 5.11). Since short-day treatments increase the nocturnal CO_2

fixation and malate synthesis of *K. blossfeldiana* (see Fig. 5.5), the above effects may be caused by decarboxylation of malate, as discussed in Chapter 5.3.2.2.

c) *Effects of Temperature*

A temperature increase during the light period will inhibit daytime CO_2 uptake. The latter is shown in Figure 5.3, which illustrates the gas exchange behavior of *Tillandsia usneoides* with respect to temperature. In this species, at temperatures below 5° C there is essentially no depression in net CO_2 uptake during the day. At temperatures above 20° C, even with saturating light intensities, net CO_2 output occurs. In principle, this pattern of gas exchange is true also for other CAM species, for instance *Kalanchoe daigremontiana* (Kluge, unpublished), and it will be shown later in detail (Chap. 6, see Fig. 6.9), that temperature-dependent net CO_2 output during the day is a feature of CAM plants in their natural habitat (in *Dudleya farinosa;* Bartholomew, 1973; in *Sedum acre* and *Sempervivum tectorum:* Kluge and Schuber, unpublished, Kluge, 1978; *Caralluma negevensis:* Lange et al., 1975). The high temperature enhancing effect on CO_2 production in the light is a reason why temperature is one of the main environmental factors affecting the in situ carbon balance of CAM plants.

The source of the released CO_2 as temperature increases is not known. There is evidence that increasing temperature causes an increase in malic acid decarboxylation (Wolf, 1960; Szarek and Ting, 1974b, see also Kluge, 1968a). If the rate of CO_2 production exceeds CO_2 refixation by photosynthesis, net CO_2 output would occur. Also, CO_2 production by mitochondrial respiration (Szarek and Ting, 1974b) and the glycolate pathway (photorespiration, see below) may also contribute to temperature-enhanced CO_2 production in the light.

Neales (1973b, 1975) showed in *Ananas comosus* and in *Agave americana* that the daytime CO_2 exchange pattern is influenced by the previous night temperature (Fig. 5.9). At constant day temperature, net CO_2 fixation in the day increased with an increase in night temperature, perhaps because nocturnal CO_2 fixation and malic acid production are inhibited by high temperatures (see Chap. 5.1.3.2,a). Hence, a lower malic acid pool during the following day results in less endogenous CO_2 production. Photosynthesis now consumes exogenous CO_2 rather than CO_2 from malic acid.

d) *Effects of Drought*

Daytime gas exchange of CAM plants tends to be greatly reduced if plants are kept under water stress; and under severe drought, no net CO_2 exchange occurs (Kluge and Fischer, 1967; Neales et al., 1968; Neales, 1975; see Fig. 5.1). Also, CAM plants growing in their natural habitat have a suppressed daytime CO_2 uptake under water stress (Bartholomew, 1973; Szarek et al., 1973; for more details see Chap. 6.2.2). Because the nocturnal CO_2 fixation is less suppressed by water deficits than CO_2 fixation in the light, drought increases the dependence of carbon gain from dark CO_2 fixation (see Fig. 6.1). This is supported by evidence of a shift in $\delta^{13}C$ values to less negative when CAM plants are kept under water stress (cf. Osmond, 1975, and Chap. 6.2.2.2).

The exact causes of the negative effects of drought on CO_2 fixation during the day remain to be studied in more detail. However, it seems reasonable to postulate that the stomata are at least partially responsible for the effect. Stomata tend to close as the plant water potential becomes more negative (see Chap. 5.3.4). It seems most unlikely that water stress sufficient to inhibit opening of stomata would have a dramatic effect on photosynthesis itself, thus inhibiting daytime CO_2 uptake.

5.1.4.3 The Initial Burst of CO_2 Uptake

Under both laboratory and field conditions, a burst of CO_2 uptake occurs immediately after the onset of illumination (see Chap. 5.1.3.1 and Fig. 5.1). Normally, this initial burst reaches its maximum within 30–60 min after illumination. Subsequently, the CO_2 exchange curve declines steeply towards the compensation point.

The temporal extension and the amount of CO_2 consumed during the burst is considerably modulated by environmental factors. In *Kalanchoe daigremontiana*, it tends to be reduced under short-day conditions (Marcelle, 1975; Kluge, unpublished); and suppressed by water stress (cf. Fig. 6.1), by high temperatures during the day, and by lowered light intensities (Kluge et al., 1973; Nobel, 1976). In contrast, the initial burst is enhanced if malic acid accumulation during the previous night was inhibited by the presence of CO_2-free air (Kluge, 1968 b), or if low temperature was applied during the day (Kluge et al., 1973). It is also enhanced under long-day conditions (Marcelle, 1975).

An unequivocal explanation of the initial burst cannot yet be provided. It has been observed by Kluge (1968 a) that CO_2 produced from malic acid stored during the previous night is utilized after the onset of illumination only after a lag phase of about 30–60 min. The lag phase corresponds with the time needed to reach the maximum CO_2 burst. Hence, it has been concluded that during the lag phase of malic acid utilization, photosynthesis depends on ambient CO_2. Only later does malic acid-derived CO_2 become available for photosynthesis. However, this is only a partial explanation, because the basic question about the cause of the malic acid lag phase is unresolved. This problem is discussed elsewhere (see Chap. 3.1.7).

In some very detailed experiments, Marcelle (1975) showed that the lag phase of malic acid decarboxylation occurred only on long-day conditions. This finding could explain the enhancing effect of long-day (i.e., short-night) treatment on the initial burst of CO_2 mentioned earlier. Marcelle (1975) postulated that the absence of the initial lag phase of malic acid utilization occurring under short-night conditions might be the result of the lack of saturation of the malic acid storage pool at the onset of light. In fact, there is evidence that under short-night conditions (however, not under long-night) the malic acid level of the tissue is still increasing during the first 60 min of the light period (Kluge, 1968 b; Marcelle, 1975; Osmond and Allaway, 1974). Osmond and Allaway (1974) showed that $^{14}CO_2$ applied exogenously during the time of the burst was fixed mainly into malic acid and remained in the malic acid pool if the pulse of $^{14}CO_2$ was followed by a chase of normal air. Hence, it seems that under short-night conditions, both

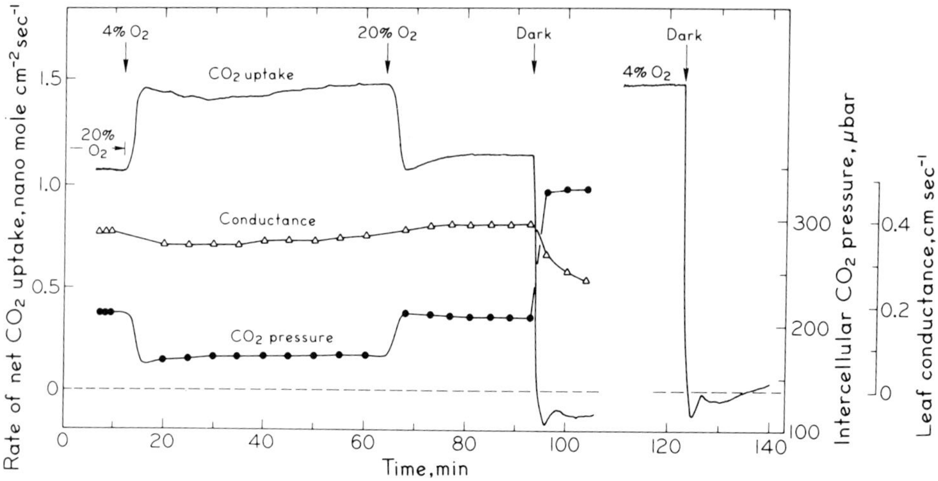

Fig. 5.12. *Kalanchoe daigremontiana:* time course of CO_2 fixation, stomatal conductance and calculated intercellular CO_2 concentration in response to ambient oxygen concentrations and darkness (leaf temperature 29° C, saturating light intensity; from Björkman and Osmond, 1974, by permission)

the C_4 pathway with concomittant malic acid synthesis and the Calvin cycle may contribute to the CO_2 fixation during the initial burst.

5.1.4.4 Compensation Point, Effects of CO_2 and O_2 Concentration on CO_2 Fixation in the Light

Estimations of compensation points in CAM plants are difficult because there is scarcely a time during the course of the diurnal cycle when gas exchange approaches steady state. More than this, it has been observed by Jones and Mansfield (1970, 1972; Jones, 1973) that shoots of *Bryophyllum fedtschenkoi*, maintained in a closed system in continuous light, have an endogenous circadian rhythm of CO_2 compensation (see also Chap. 5.1.5). Allaway et al. (1974a) suspected that the same might be true in *Kalanchoe daigremontiana.* Jones and Mansfield (1972) showed that the rhythm of CO_2 compensation was susceptible to changes in temperature and light intensity. For instance, a light intensity of 78 J $m^{-2}s^{-1}$ was sufficient to damp the endogenous rhythm at 15°, but not at 10° C.

The minimum endogenous oscillation in CO_2 compensation was during the prospective midnight, and the maximum was during the prospective midday. During the troughs of the curve, the CO_2 concentration within the closed system was as low as 0 to 5 ppm CO_2. During the maximum, the CO_2 compensation was on the order of 60 to 100 ppm CO_2, depending on temperature and light intensity.

In *Kalanchoe daigremontiana*, Allaway et al. (1974a) estimated the CO_2 compensation point at the end of a 16-h day, when some constancy of light CO_2 fixation was approached, at 51 μl/l (=92 ng cm^{-3}) (saturating light intensity). This value is comparable to that of C_3 plants, i.e., higher than that of C_4 plants (cf. Downton, 1971).

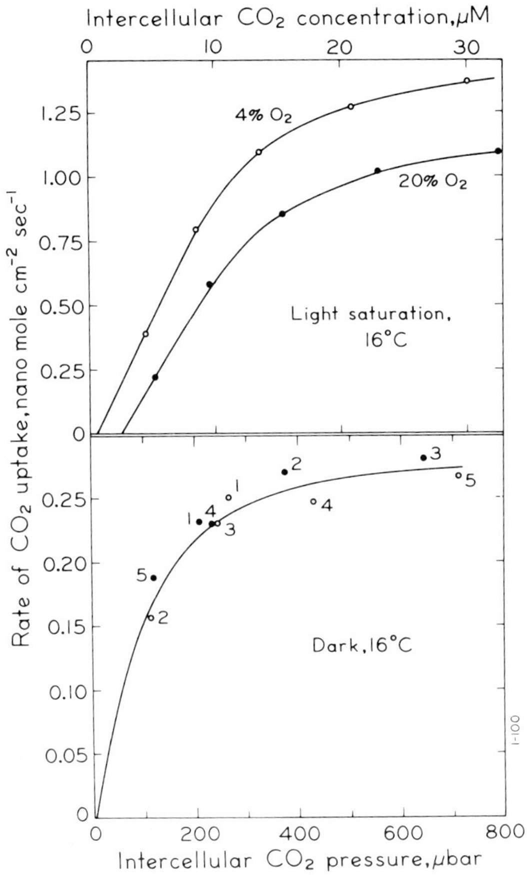

Fig. 5.13. Rate of CO_2 fixation 16° C in *K. daigremontiana* at two different oxygen concentrations at saturating light intensity (*top*) and darkness (*bottom*) depending on intercellular CO_2 pressure (from Björkman and Osmond, 1973, by permission)

Extensive studies, again with *Kalanchoe daigremontiana*, of the relationship between CO_2 and O_2 concentration and CO_2 fixation were provided by Osmond and Björkman (1975) and Björkman and Osmond (1974). These authors demonstrated that light-dependent CO_2 fixation showed a lower affinity for CO_2 (half saturation of 9 μM at 16° C and 37 μM at 29° C) than did dark CO_2 fixation (half-saturation 3 μM at 16° C). The rate of dark CO_2 fixation was not affected by changing oxygen concentrations, whereas light CO_2 fixation was considerably less in oxygen of 20% than in 4% O_2 (Fig. 5.12). The inhibitory effect of high O_2 concentrations on light CO_2 fixation occurred throughout the observed range of intercellular CO_2 pressures (up to 800 μbars; Fig. 5.13). The negative effect of high oxygen concentration on light CO_2 fixation can reasonably be explained in terms of the oxygenase reaction of RuDP carboxylase (Bowes and Ogren, 1972; Osmond, 1975; Ogren, 1975). Hence, the above results of Osmond and Björkman support the view that in the light, CAM plants fix exogenous CO_2 primarily via

RuDP carbocylase and the Calvin cycle rather than via the C_4 pathway. In contrast, the lack of oxygen inhibition of dark CO_2 fixation suggests that RuDP carboxylase is not involved in harvesting CO_2 from the atmosphere during the night.

5.1.5 CO_2 Exchange in Continuous Darkness or Continuous Light

5.1.5.1 Introduction

It has long been known that a rhythmic gas exchange in CAM plants can persist under continuous darkness (see Bünning, 1973; see Fig. 5.14). Thus, the gas exchange of certain CAM plants follows a typical circadian endogenous rhythm. The question of rhythms in CAM has previously been reviewed by Wilkins (1960, 1969) and Queiroz (1974).

According to Pittendrigh (1954; see also Wilkins, 1969), true endogenous rhythmic phenomena obey the following:

1. The rhythm persists under constant environmental conditions
2. The rhythm phase can be shifted by environmental perturbation, and the new rhythm will persist if kept under constant conditions
3. A single stimulus is sufficient to set or alter a rhythm
4. Since rhythms depend on energy (for example, oxidative phosphorylation), anaerobiosis will inhibit.

5.1.5.2 CO_2 Exchange in Continuous Darkness

a) General Phenomenology

Typically, experiments have been performed by measuring CO_2 release into CO_2-free air from detached leaves, under constant dark or light conditions. Schmitz (1951) first reported such rhythmic CO_2 output patterns from excised leaves of *Kalanchoe blossfeldiana* in continuous darkness. *In Bryophyllum fedtschenkoi*, Wilkins (1959, 1962a, b, 1967) studied a CO_2 rhythmic phenomenon which he termed simply "CO_2 output." Basically, he found a rhythmic output of CO_2 into CO_2 free air. In continuous darkness, at the time of the normal night cycle, the lowest CO_2 output was measured, whereas during the normal day, the CO_2 output was higher (Fig. 5.14). Curves similar to those of Figure 5.14 have been interpreted by Wilkins as a manifestation of an endogenous rhythm of CO_2 output. However, CO_2 analyzers such as those used by Wilkins measure only net CO_2 exchange and cannot differentiate between rates of output and uptake. Hence, it is not entirely clear if the rhythmic CO_2 output such as Wilkins described is a true output rhythm or the result of changing uptake in the presence of constant output. Dark CO_2 fixation in the presence of CO_2 release would complicate interpretation.

In order to clarify this question, Warren and Wilkins (1961) supplied $^{14}CO_2$ in continuous darkness to leaves of *Bryophyllum fedtschenkoi* and measured simultaneously total net CO_2 exchange. The results of this experiment suggested that it was the $^{14}CO_2$ uptake that showed the rhythm. The $^{14}CO_2$ fixation maximum

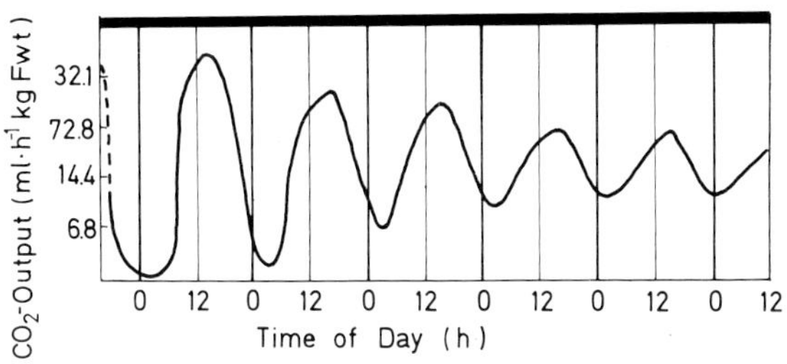

Fig. 5.14. CO_2 output rhythm in continous darkness (detached leaves of *Bryophyllum calycinum*; after Bünning, 1963; redrawn from Mohr, 1969, by permission)

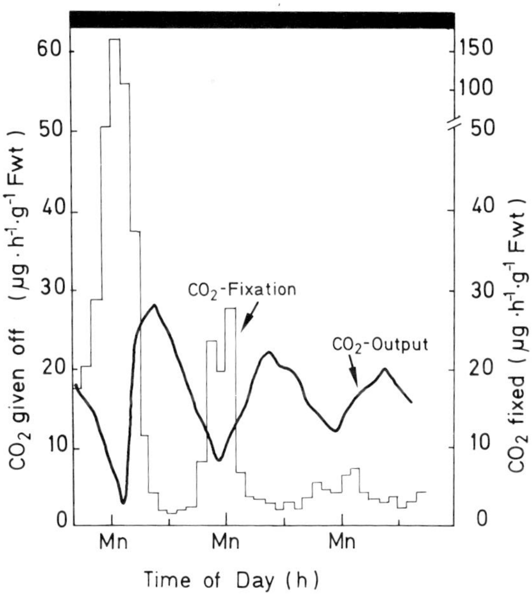

Fig. 5.15. Rhythms of CO_2 output and $^{14}CO_2$ fixation in leaves of *Bryophyllum fedtschenkoi* under continuous darkness. *Solid line:* CO_2 output; *histogram:* CO_2 fixation (after Warren and Wilkins, 1961, by permission)

corresponded exactly with the minimum of net CO_2 output, and vice versa (Fig. 5.15). This coincidence of dark CO_2 fixation phase and CO_2 output rhythm could not be separated by experimental treatments. For example, the onset of continuous dark at different times of the day/night cycle and the pretreatment by short or long days did not alter the pattern. Hence, Warren and Wilkins (1961) suspected that the rhythm of CO_2 output was in fact a rhythm of dark CO_2 fixation.

Assuming that the dark release of CO_2 comes from oxidative respiration, continuous O_2 measurement should clarify the nature of the CO_2 exchange rhythm. Until such experiments have been performed and interpreted, the neutral term CO_2 output to refer to the rhythm is probably more appropriate.

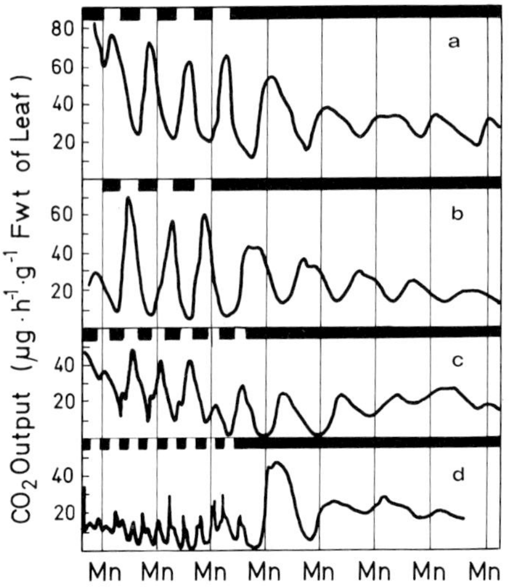

Fig. 5.16

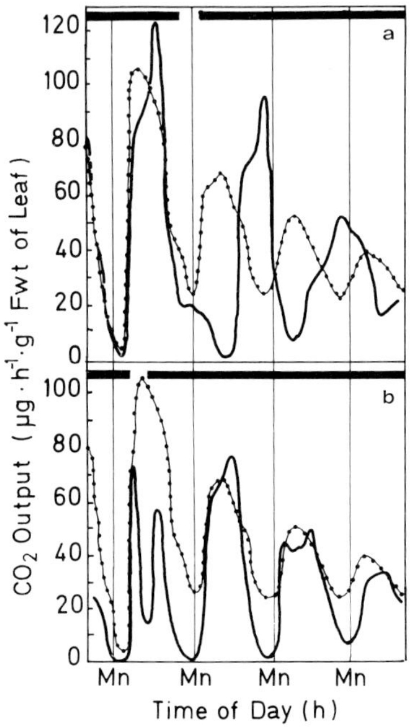

Fig. 5.17

Fig. 5.16. The apparent entrainment of CO_2 output rhythm by exposing leaves of *Bryophyllum fedtschenkoi* to 8:8 (**a**, **b**), 6::6 (**c**) and 3:3 h cycles of light (1000 lux/darkness). The illumination programme is indicated by the bar above each graph. *Mn*, midnight (after Wilkins, 1960, by permission)

Fig. 5.17 a and b. The effect of interruption of continuous darkness by short light treatments on the CO_2 output rhythm in *Bryophyllum fedtschenkoi*. **a** Light treatment during the trough of CO_2 output curve. **b** Light treatment during the peak of CO_2 output curve. ●–●–● Control remaining under continuous darkness. —— light-treated sample *Mn*, midnight (after Wilkins, 1969, by permission)

b) Effects of External Stimuli

The rhythm of CO_2 output in *Bryophyllum fedtschenkoi* can be altered by light interruption of continuous darkness (Wilkins, 1960). Entrainment occurs by periodic cycles of weak light (1000 lux) of 8:8, 6:6, or 3:3 h (Fig. 5.16). Then, after the onset of continuous dark once again, there is an immediate reestablishment of the approximate 23-h circadian rhythm. The phase of CO_2 output can be reset by a single 3-h light interruption of continuous darkness, providing the CO_2 output cycle is in the proper position (Fig. 5.17). The phase of rhythm is only affected if the light stimuli occur during the CO_2 output minimum period (Wilkins, 1960). Red light, but not blue, acts similarly to white light (Wilkins, 1960).

c) Inhibition of the CO_2 Output Rhythm

The CO_2 output rhythm of *Bryophyllum fedtschenkoi* can be inhibited by interrupting continuous dark by bright light or high temperature (Wilkins, 1959, 1960, 1962a, b). After light or temperature treatment, restoration of the rhythmic oscillation can be brought about by reestablishment of the normal dark or temperature conditions. Based on these observations, Wilkins (1969) concluded that both bright light and high temperature effect the rhythm by setting a basic oscillator to the same phase of the cycle, and holding it there. The details of such a mechanism are, of course, largely unknown.

Because general metabolic activity is strongly inhibited by either low temperature or anaerobiosis, it is not surprising that these treatments inhibit the CO_2 output rhythms (Wilkins, 1962a, 1967). The application of low temperature or anaerobiosis for a limited interval causes phase shifts. Similarly to light (see Fig. 5.17), the effect depends on the time of the cycle when the treatments are applied (Wilkins, 1962a, b, 1967).

d) CO_2 Output Rhythms in Normal Air

It is not known if the CO_2 output rhythms described above are accompanied by corresponding oscillations in malic acid concentrations. Clearly, such acid changes would occur if the rhythmic phenomena were CAM-related. Since CO_2-free air inhibits malic acid accumulation even in normal day/night cycles (Wolf, 1960; Kluge, 1968b), the rhythmic phenomena described by Wilkins may not be CAM-related. However, Queiroz (1975) reported a circadian CO_2 output rhythm by *Kalanchoe blossfeldiana* in normal air in darkness as did Wilkins (1959) with *B. fedtschenkoi*. Hence, the rhythms are probably not the result of the CO_2-free air, but since we do not know more about corresponding malic acid fluctuations, we cannot determine the extent to which the rhythmic CO_2 output is CAM related.

5.1.5.3 CO_2 Exchange Under Continuous Illumination

There are some studies, though less extensive, of CO_2 rhythms in constant light. Nuernbergk (1961) reported a circadian oscillation of net CO_2 uptake by *Bryophyllum daigremontianum* and other CAM species under constant illumination and in normal air. Queiroz (1975) also reported a circadian rhythm of CO_2 exchange by *Kalanchoe blossfeldiana* in constant light and normal air which dampened after two 24-h periods. In continuous dark, the rhythm persists for an extended period. In general, the continuous light CO_2 uptake rhythm shows a minimum during the day period and a maximum during the period that would normally be dark (Fig. 5.18).

In *Kalanchoe tubiflora*, the light CO_2 uptake can be shifted to CO_2 output by high temperature (35° C). After resetting the lower, normal temperature (20° C), the circadian rhythm phase of CO_2 uptake remains shifted 180° with respect to the control (Fig. 5.18). Furthermore, under constant light and temperature, the phase shifts slowly, such that the maximum CO_2 uptake occurs during the normal day period.

In *Kalanchoe tubiflora*, during the first cycle the CO_2 uptake rhythm in continuous light and normal air is accompanied by a predictable oscillation in malic acid (Kluge, 1969c). The malic acid concentration increases with CO_2 uptake and decreases during the low CO_2 uptake period. In *Bryophyllum calycinum*, Kress in Kluge's laboratory (unpublished) showed that the rhythm of CO_2 uptake and

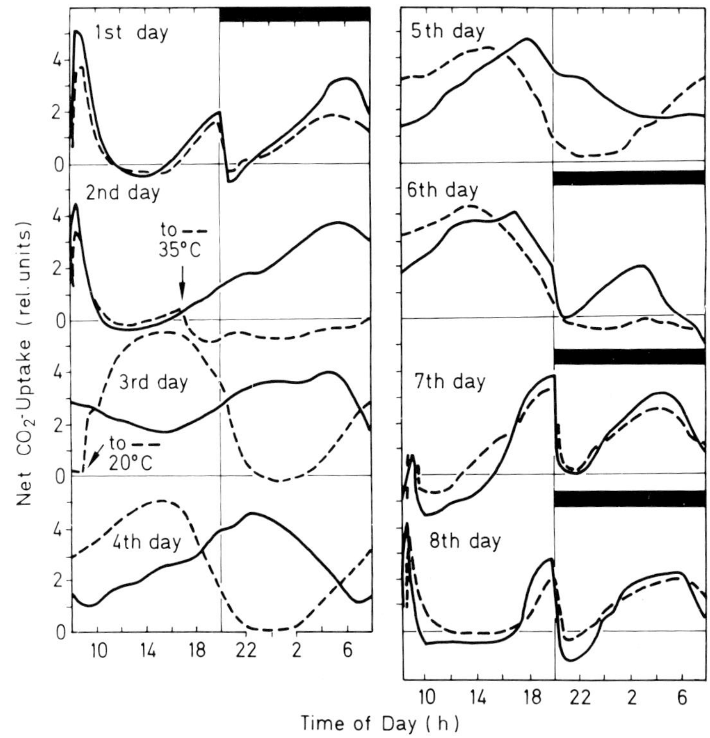

Fig. 5.18. Rhythm of CO_2 exchange performed by Kalanchoe tubiflora in normal air. After precultivation of the plants under 12:12 h light/dark rhythm, two individuals were transferred to four 24 h periods of continuous illumination (10,000 lux). One of the plants was treated on the second day at 18 h with 35° C. A phase shift of CO_2 exchange occurred after this treatment

malic acid oscillation persists for at least two weeks when in constant light. However, the oscillations of the malic acid rhythms were considerably damped. Also Lüttge (unpublished) found that the oscillation of malic acid level damps faster than the rhythm of CO_2 exchange.

5.1.5.4 Conclusions

Overall, it seems reasonable to conclude that CAM CO_2 exchange is a good example of endogenous rhythmic phenomena meeting all the requirements of Pittendrigh (1954). Frequently, the occurrence of endogenous rhythms is interpreted as the manifestation of a "biological clock" (Bünning, 1973; Sweeny, 1969; Queiroz, 1974). Since CAM is an example of a thoroughly understood rhythmic phenomenon, future subsequent work could lead to a causal explanation of an endogenous rhythm.

Such speculation presumes, however, that the rhythm is a general feature of CAM plants. Despite the fact that *Kalanchoe blossfeldiana*, *K. daigremontiana*, and *Bryophyllum fedtschenkoi* are typical CAM plants, only a few species have been studied precluding generalization. In fact, *Kalanchoe tubiflora*, a very typical CAM plant, only shows a CO_2 rhythm in continuous light and not dark (Wilkins, 1960). Nuernbergk (1961) reported that succulent orchids which showed dark CO_2 fixation under normal day/night cycles lack rhythmic CO_2 exchange in continuous illumination and normal air. Finally, as indicated above, the CO_2 exchange rhythms may persist without malic acid oscillation.

Hence, despite much information concerning the CAM rhythmic phenomena, an unequivocal model is lacking. We do not know, for example, if the CO_2 rhythm is the result of changing CO_2 production capacity or a rhythmic alteration in carboxylation capacity. Until now, only a dark CO_2 fixation rhythm is established.

In a series of extensive and detailed studies, Queiroz has reported much evidence that both CO_2 production and fixation are rhythmic. The circadian rhythm of CO_2 exchange is thus explained by Queiroz (1974) in terms of endogenous rhythms of enzyme activity (cf. Chap. 4.2.3.2,b). Queiroz's data on enzyme capacity indicate that the rhythm of CO_2 fixation is an overt indication of actual P-enolpyruvate carboxylase activity which oscillates synchronously with CO_2 fixation (Queiroz, 1974). Similarly, the rhythm of CO_2 production follows the enzymic capacity of the endogenously controlled malate enzyme (see Fig. 4.6 for the rhythmic phases of PEP carboxylase, malate enzyme, and malic acid).

We do not, of course, know about the nature of the basic oscillator (biological clock!) which finally regulates the enzymic capacity rhythms, and hence, the enzyme rhythm model is still incomplete.

As outlined earlier (see Chap. 4.2.3.2), Lüttge in cooperation with Kluge (Kluge and Heininger, 1973; Lüttge and Ball, 1974; Lüttge et al., 1975, 1977) proposed a model which avoids the biological clock dilemma. This model is based on the assumption that the malic acid storage capacity of the vacuole regulates the malate level of the cytoplasm via regulation of malate fluxes across the tonoplast (see also Kluge, 1971 b, 1976 b, 1977 a). The cytoplasmic malate pool, then, could regulate in a "feedback loop" PEP carboxylase and hence, CO_2 fixation (see Chap. 4.2.3.2). The crucial feature of this model is the assumption of a turgor mechanism which causes the release of malic acid (malate) out of the vacuole into the cytoplasm. As outlined by Lüttge et al. (1975), this mechanism would be similar to the tension–relaxation postulate of Bünning (1973) operating as a basic oscillator in biological clocks. In the above model of Lüttge, the tension phase is represented by vacuolar filling with malic acid and the relaxation phase corresponds to the loss of malic acid from the vacuole to the cytoplasm.

Although this model is as yet tentative, it allows explanation of how environmental stimuli could alter the rhythmic behavior of CAM. For example, the dampening effect of light (especially red) on the CO_2 output rhythm in continuous darkness may be due to a light-mediated alteration in vacuole permeability. The assumption of a light-altered permeability of the tonoplast resulting in a decreased capacity for malic acid storage has recently been suggested by Nalborczyk et al. (1975) who observed substantial CO_2 production by malate decarboxylation after far red or white light illumination of *Kalanchoe daigremontiana* leaves.

High temperature dampening of CO_2 exchange and malic acid rhythms may result because of the prevention of vacuole accumulation of acid, and thus a prevention of resetting the rhythm. These external factors, therefore, would prevent the tension portion of the Bünning model from destroying its oscillatory nature.

5.2 Oxygen Exchange

5.2.1 History

Because of the extreme difficulty in measuring small changes in O_2 in an atmosphere of nearly 21% oxygen, studies of oxygen exchange have lagged far behind those of CO_2 exchange. The first reported observations of O_2 exchange in a CAM plant *(Opuntia)* were those of De Saussure (cited in Wolf, 1960; see footnote 2 on p. 108).

Using manometric analyses, a variety of researchers measured respiratory gas trends of succulents (Wolf, 1938; Thomas and Beevers, 1949; Thomas and Ranson, 1954; Moyse, 1955a, b; Moyse and Jolchine, 1956; Bruinsma, 1958). These studies showed that during the dark acidification phase (when CO_2 is being taken up), respiratory quotients (CO_2/O_2) may be negative.

5.2.2 Manometric Analysis

Recently, Szarek and Ting (1974b) studied O_2 exchange in *Opuntia* by manometric means. Consistent with previous CO_2 exchange studies (see Tables 5.2, 5.3), O_2 consumption increased with increasing temperatures. Gas exchange in general was a function of the age and physiological condition of the tissue. For example, young stem sections had greater respiratory activity than old, mature sections.

Stem sections collected at the end of the light period, when malic acid was low, showed O_2 consumption in light or dark. Those sampled at the end of the dark period, when acid levels were high, showed a low rate of O_2 evolution in the light indicative of some measurable photosynthesis.

5.2.3 Polarographic Analysis

Denius and Homann (1972) studying *Aloe arborescens* Mill. with polarographic methods showed that the decrease of malic acid in the light was accompanied by O_2 evolution. Both the malic acid consumption and O_2 evolution were inhibited by amytal, cyanide, trifluoroacetate, and 2-dimethyl malonate, suggesting to them that respiration may be involved in the light-dependent decarboxylation of malate. The elaborate measurements of oxygen exchange in *K. daigremontiana* by Kaplan et al. (1976a, b, 1977) are discussed elsewhere (Chaps. 5.1.3.2,a and 5.1.3.2,b).

5.2.4 Paramagnetic Analysis

Paramagnetic oxygen analyzers (Firma Mayhac, Hamburg) allowed Brunnhöfer et al. (1968a, b) and Marcelle (1975) to measure oxygen exchange of whole CAM plants continuously over complete diurnal cycles. These authors showed

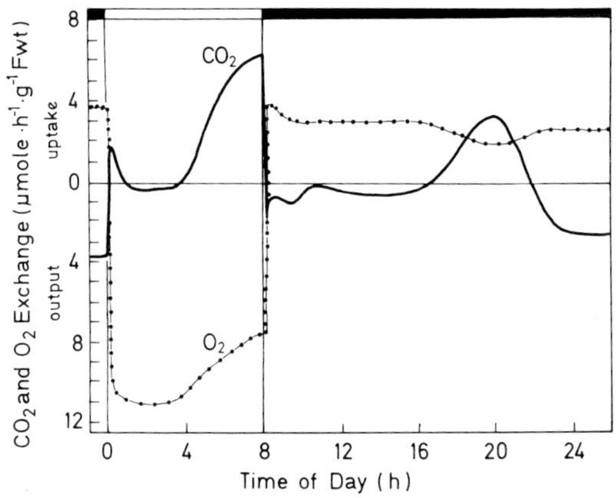

Fig. 5.19. Diurnal pattern of CO_2 (——) and O_2 (······) exchange in detached leaves of *Kalanchoe daigremontiana*, temperature 20° C (modified after Brunnhöfer et al., 1968a)

that oxygen production in the light was decreased and consumption was increased as night temperature was raised from 10° to 30° C (see also Szarek and Ting, 1974). Marcelle (1975) studied the pattern of oxygen (and CO_2) exchange of *Kalanchoe daigremontiana* grown under short- and long-day conditions. Oxygen was evolved in the light with maximum rates during the depression of CO_2 uptake (see Fig. 5.6 and 5.19). Attempting to explain the light curves of oxygen exchange, it is reasonable to assume that the high O_2 evolution when CO_2 uptake was nil might reflect photosynthetic consumption of CO_2 supplied from the endogenous storage pool, i.e., malic acid. The decrease in O_2 output typical for the last hours of the light period during the time when CO_2 uptake was increasing could be explained, assuming that exogenous CO_2 is not only fixed by the reductive pentose phosphate cycle (thus not only linked to oxygen production). This assumption is supported by the fact that towards the end of the light period substantial $^{14}CO_2$ is fixed into malic acid rather than sugar phosphates (Kluge, 1969c, 1971b).

In summary, the oxygen exchange of CAM plants is quite easily modified by a variety of exogenous factors such as temperature, light intensity, and day length, and by endogenous factors such as tissue age, malic acid level, and the period of the diurnal CAM cycle. Similarly to CO_2, there is little steady-state O_2 evolution in the light except perhaps for extended day periods under high light intensity. Because of this apparent lack of a steady-state period, studies of the relationship between internal and external factors are difficult to interpret.

5.3 Water Vapor Exchange and Stomata of CAM Plants

5.3.1 Introduction

Gas exchange processes of plants, in particular exchange of water vapor (transpiration), is to a large extent mediated by the stomata. Transpiration can be considered a physical process depending on vapor pressure gradients between the plant and the surrounding environment and on the characteristics of the diffusion pathways.

The vapor pressure gradient is solely a physical property depending upon environmental factors. In contrast, the stomata with variable apertures respond to plant parameters as well.

Provided that the magnitude of cuticular transpiration is insignificant (which is true in the majority of CAM plants), the transpiration rate is a convenient indicator of the diffusive capacity of stomata. Hence, it is reasonable to consider the properties of transpiration as observed in CAM plants in terms of stomatal movements, and vice versa. In this chapter, we will therefore mainly detail the current knowledge of stomata in CAM plants. The ecological implications of water vapor exchange of CAM plants will be discussed in detail in Chapter 6.

5.3.2 The Diurnal Cycle of Stomata Movements

5.3.2.1 Phenomenology

By far the most interesting and significant feature of CAM plant stomatal behavior is that they may open at night and close during the day.

As early as 1907, Livingston (1907) studied in detail transpiration of Arizona cacti relative to evaporation recorded that, unlike other nonsucculent plants, relative transpiration was high at night and low during the day. Livingston concluded from his data that "... there exists in the cacti a mechanism for governing the rate of water loss, which is entirely different in its response to external conditions or in its daily periodicity from the corresponding mechanism in leafy plants." Shreve (1916) working at the Carnegie Desert Laboratory at Tucson found that the stomata of cacti were more open at night than in the day, thus accounting for the greater night transpiration in cacti despite a much greater evaporative demand during the day.

Although the early work of De Saussure (see Chap. 5.1.1) clearly showed dark CO_2 uptake by cacti (see Chap. 5.1.1), it was not until the early reports of Gregory et al. (1954) and Nuernbergk (1961) showing nocturnal rather than daytime CO_2 uptake that the significance of the reverse phase gas exchange became of general interest (see Chap. 6).

Subsequently, reports such as Nishida (1963, see also Nishida, 1977) and Joshi et al. (1965) confirmed and emphasized nocturnal stomatal opening in succulents.

Following the classification of CAM gas exchange patterns as suggested by Neales (1975) and discussed above (see Chap. 5.1.2; see Fig. 5.1), three types of transpiration patterns in CAM plants (i.e., three types of stomatal movement patterns) can be distinguished:

a) Stomata open towards the end of the night and remain open during the entire day

b) Stomata are open during most of the night. Occasionally there is an increase in aperture during the time of the initial CO_2 burst (see Fig. 5.1), followed by stomatal closure. The stomata open again at the end of the light period (Figs. 5.1, 5.10, 5.20)

c) Stomata are open during the night and closed during the entire light period (Fig. 5.21).

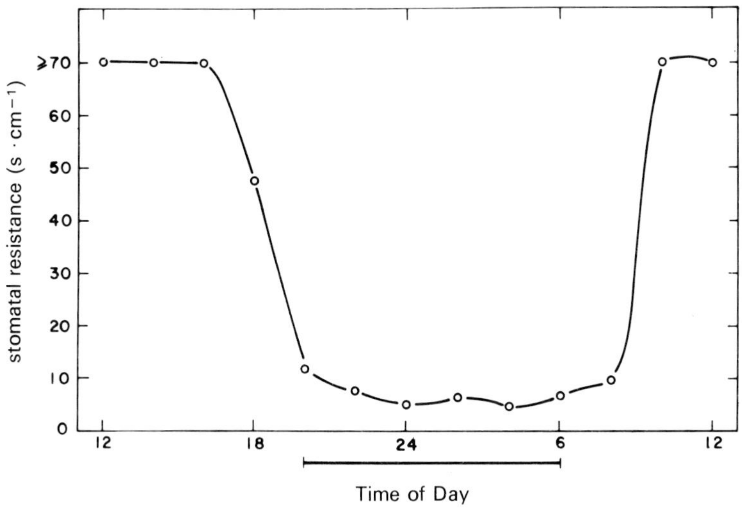

Fig. 5.20. Diurnal pattern of stomatal resistance for *Opuntia basilaris* growing in southwestern United States at the Boyd Desert Laboratory. The stomata are open at night and closed during the day. Estimates were made with a porometer after precipitation

Similar to the CO_2 exchange pattern (see Chap. 5.1.2), any given plant may show variations in the basic pattern of stomatal behavior mainly in response to environmental factors (see Chap. 5.3.4).

Figure 5.20 depicts a typical diurnal pattern of stomatal resistance for *Opuntia* growing in the deserts of the southwestern United States. Leaf resistance is high during the day-light period. Toward the end of the day, stomata begin to open, resulting in reduced gas transfer resistances. Throughout the night period, stomata are open and gas transfer resistances are low.

Crassulacean succulents, e.g., *Kalanchoe daigremontiana*, have much the same pattern, but frequently show stomatal opening in the afternoon (Fig. 5.21). The afternoon opening could be the result of a reduced partial pressure of CO_2 after organic acid depletion and photosynthetic reduction of endogenously produced CO_2 as outlined in Chapter 5.3.2.2.

5.3.2.2 Coupling Between CAM and Movements of Stomata

As discussed above in detail, the unusual pattern of CO_2 exchange of CAM plants, with nocturnal net CO_2 fixation, is accompanied by an equally unusually pattern of stomatal movement, with stomata being open during the night and closed during the day. It has been shown by Nishida (1963), Kluge and Fischer (1967), Neales (1975), Kristen (1969), Osmond and Allaway (1974), and others that there is a close relationship between CO_2 uptake and stomatal opening. The CO_2 exchange of CAM plants follows exactly the course of stomata with net CO_2 uptake occurring when stomata are open (Fig. 5.22). Because of the clear coincidence between stomatal movement and CO_2 exchange, it is tempting to conclude that stomatal behavior is the primary event determining CO_2 exchange

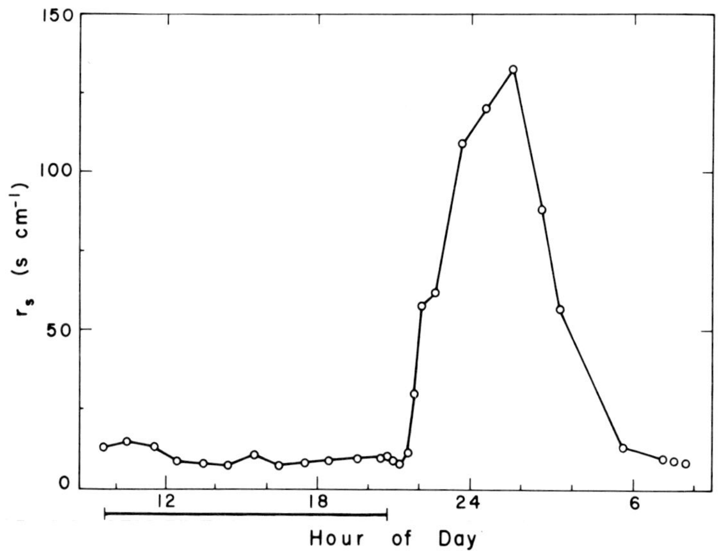

Fig. 5.21. Diurnal patterns of stomatal resistance for an experimental *Kalanchoe daigremontiana* plant grown in a pot. The pattern is similar to *Opuntia basilaris* except for opening at the end of the light period. Data were obtained by calculation from transpiration estimates by Osmond. The time along the abscissa is separately experimental

in CAM plants. However, Kluge and Fischer (1967; see Fig. 5.22) reported data showing the opposite. They postulated that CO_2 exchange of the mesophyll cells, which itself is the direct consequence of CAM, determines whether the stomata close or open at a given time.

CAM and stomatal movement might be coupled by CO_2 concentrations in the substomatal intercellular spaces (see Raschke, 1966, 1975a). Low internal CO_2 concentrations can be expected to occur during the night because of effective dark CO_2 fixation by PEP-carboxylase. Thus, the low CO_2 internal concentration will cause the stomata to open. During the day, because of malate decarboxylation the internal CO_2 concentration will increase, causing the stomata to close. In fact it has been shown by Kluge (1968b, see Fig. 5.10) that stomatal closure during the day indicated by the depression of net CO_2 uptake lasts only as long as malate is consumed in the light.

The above postulate of Kluge and Fischer (1967) is founded on experiments with *Kalanchoe daigremontiana* where both transpiration (i.e., stomatal movement) and CO_2 exchange were measured before and after the stomatal-bearing epidermis had been removed from the leaves. After removal of the stomata, the transpiration lost its rhythm, formerly coincident with the rhythm of CO_2 exchange (see Fig. 5.22). The rhythm of CO_2 exchange, however, remained in principle unchanged. Furthermore, strong support comes from results obtained by Neales (1970), who studied the effects of various ambient CO_2 concentrations on the stomatal behavior of *Agave americana*. Again, high CO_2 concentrations tended to close the stomata, low concentrations, however, having the opposite effect.

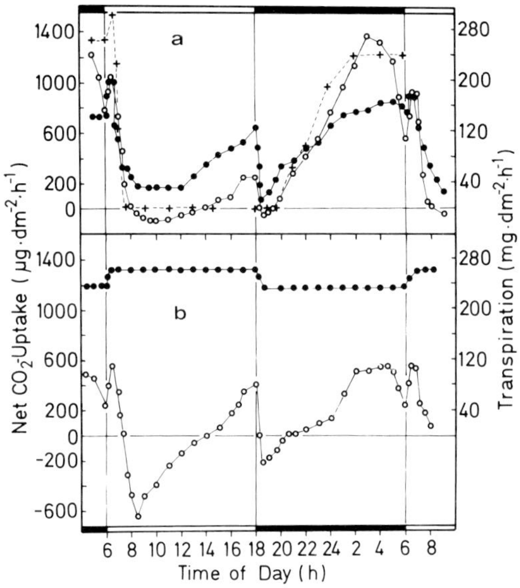

Fig. 5.22 a and b. Net CO_2 exchange and transpiration in *Kalanchoe daigremontiana*. **a** Coincidence of CO_2 exchange and transpiration in an intact plant. ○—○ net CO_2 exchange, ●—● transpiration, ×—× porometric values of stomatal movement obtained by Nishida (1963). **b** CO_2 exchange and transpiration in the same plant as **a**, however, after the lower epidermis of all leaves has been removed (after Kluge and Fischer, 1967)

The postulate of a causal linkage between CAM-determined CO_2 exchange of the mesophyll cells and stomatal movement has a general aspect. It implies that the transpiration pattern (determining the water balance of the plant) is a direct consequence of the specific mode of photosynthesis called CAM. It is this close causal connection between carbon and water balance which provides the ecological advantage in situ as discussed subsequently (Chap. 6).

5.3.2.3 Mechanism of Stomatal Opening

The mechanism of stomatal opening has recently been reviewed by Raschke (1975a, 1976). Basically the physical driving force for stomatal opening is increased turgor of the guard cells relative to adjacent cells, which causes expansion and subsequent opening (Raschke, 1975a). The increased turgor comes about by an energy-dependent accumulation of solutes, the osmoticum for water uptake causing the hydrostatic pressure or turgor. It is known that CO_2 can regulate stomatal aperture. Increasing CO_2 results in a decreased aperture, although high water status may cause a decreased CO_2 sensitivity. Abscissic acid is necessary for CO_2 sensitivity (Raschke, 1975b).

Light may act as a signal for stomatal opening, but Raschke interprets this to mean that light reduces CO_2 and it is the CO_2 and not light which regulates stomatal opening. Light in fact, does not appear necessary for stomatal opening (Raschke, 1975a).

Meidner and Mansfield (1968), in their book on stomatal physiology, indicated that there was no evidence that the mechanism for night opening of CAM plant stomata was any different from day opening of other plant stomata. Given the present literature (and see Raschke's reviews, 1975, 1976) the latter seems to be a reasonable conclusion. Day opening of stomata is a response to reduced CO_2 because of photosynthesis, whereas in succulent plants with nocturnal opening, the reduced CO_2 comes about by dark CO_2 fixation accompanying the CAM metabolism (see Chap. 6.3.2.3).

Recent investigations have indicated that the osmotica creating reduced osmotic potentials and subsequent water uptake by guard cells are potassium and endogenously synthesized organic anions. Studies by Fujino (1967, 1969) and Fischer (1968) among others have shown K^+ accumulation in guard cells, and Allaway (1973) reported malate sufficient to balance about 50% of the K^+. Other organic acids such as citrate were also present. The fact that guard cells fix CO_2 in the dark (Nishida, 1963) and have PEP carboxylase (Willmer and Pallas, 1973) has led to the conclusion that PEP carboxylase is involved in organic acid synthesis during stomatal opening.

Significant for the understanding of succulent CAM plant stomatal opening is that Dayanandan and Kaufman (1975) reported K^+ accumulation in the guard cells of *Crassula argentea* during night opening.

As indicated earlier, we conclude that the night stomatal opening of CAM plants is initiated by reduced partial pressures of CO_2, because of dark CO_2 fixation. Potassium, malate, and perhaps other organic anions accumulate and reduce osmotic potentials, resulting in water uptake, increased turgor, and stomatal opening. The mechanism is the same, therefore, as in nonsucculent plants. In the light, CO_2 is kept high by malate decarboxylation (Chap. 3). The observation that succulent plant stomata frequently open toward the end of the light period is consistent with this hypothesis.

5.3.3 Gas Diffusion Resistances in CAM Plants

5.3.3.1 Gas Exchange Parameters

There are many discussions in the literature of the gas exchange parameters of plants (see, e.g., Slatyer, 1967). Transpiration or photosynthesis can be described by analogy with Ohm's law by the first Fick's law as follows (Penman and Schofield, 1951):

$$Q = \Delta / r \tag{5.1}$$

where

Q = gas taken up or lost, i.e., photosynthesis or transpiration in g cm^{-2} s^{-1},

Δ = the gas concentration (either CO_2 or H_2O) between the source and sink in g cm^{-3},

and

r = resistance to gas transfer in units of s cm^{-1}.

Frequently, CO_2 uptake or water loss are reported in units of g dm^{-2} h^{-1} rather than g cm^{-2} s^{-1}.

The r term of Eq. (5.1) may be factored into component resistances in series or parallel. For example, for transpiration, r can be factored into r_a and r_s, where
r_a = the boundary layer or air resistance, and
r_s = resistance accountable for by number, size, and spacing of the stomata. r_s is a variable resistance.

In addition, the resistance accounted for by the nonpored surface of the leaf can be included. This resistance, termed cuticular resistance, r_c, is in parallel with r_s and can not be added directly. Since it is usually large and constant, for most purposes it can be ignored because it will not significantly alter the sum of $r_a + r_s$.

For CO_2 uptake, an additional in series resistance, r_m is present and frequently termed the residual resistance which includes CO_2 transport and the biochemistry of carboxylation. r_a and r_s can be converted from H_2O to CO_2 by the factor 1.56.

5.3.3.2 Boundary Layer Resistance (r_a)

In general, boundary layer considerations have centered around water vapor or heat transfer from flat surfaces comparable to many leaves. Succulents, however, tend toward a cylindrical or spherical shape, and perhaps the ideal succulent form minimizing surface area to volume would be a perfect sphere (Ting and Szarek, 1975, cf. also Chap. 2.1). Considerations by Powell (1940) and more recently for plants by Nobel (1974, 1975) have shown that evaporation from a spherical surface or conversely, the resistance to evaporation (i.e., the boundary layer resistance), is an exponential function of wind velocity and effective direction. Calculations using Powell's (1940) expression,

$$E_s = 2.1 \cdot 10^{-7}\,(\mathrm{VD})^{0.59}\,D^{-1} \tag{5.2}$$

where
E_s = evaporation from a spherical surface,
V = wind velocity, and
D = diameter
indicate that the boundary layer resistance of a sphere is greater than a disc of comparable area (Ting and Szarek, 1975).

Although we know of no specific measurements of r_a for succulents, Nobel (1975) estimated r_a for near spherical fruiting bodies of Basidiomycetes.

For wind of 20–200 cm s^{-1} and diameters of 1.4–7.9 cm, estimated r_a valves ranging from 0.16 to 1.29 s cm^{-1}. The 0.16 s cm^{-1} was for a *Lycoperdon* fruiting body 1.4 cm in diameter and a wind velocity of 200 cm s^{-1}, whereas the 1.29 s cm^{-1} estimate was for a *Calvatia* body 7.9 cm in diameter and 20 cm s^{-1} wind.

It seems reasonable to conclude from these studies that the r_a of succulent plants will be low on the order of 0.2–1.5 s cm^{-1}.

5.3.3.3 Stomatal Resistance (r_s)

In the literature there are many quantitative studies of the stomatal resistance of succulent plants (Ting et al., 1971). Because of few stomata and long apparent

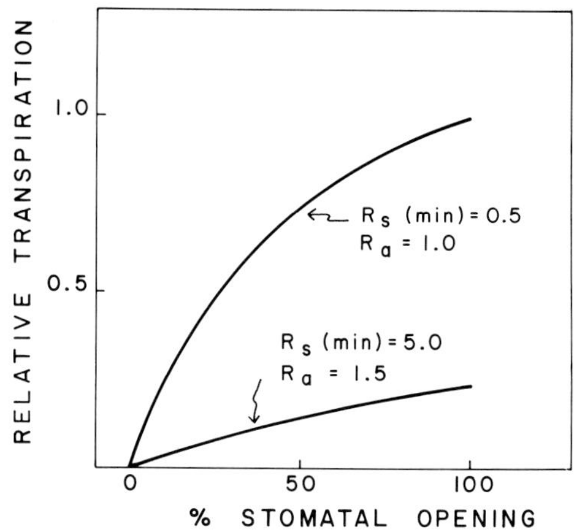

Fig. 5.23. Comparison of stomatal control of transpiration by a succulent plant with a minimum r_s of 5 s cm^{-1} and an r_a of 1.5 s cm^{-1}, with a nonsucculent with minimum r_s of 1.0 s cm^{-1} and r_a of 0.5 s cm^{-1} (from Ting and Szarek, reprinted by permission from Environm. Physiol. of Desert Organisms; N.F. Hadley e.d.; Dowden, Hutchinson and Ross Inc. Stroudsburg, 1965)

substomatal diffusion path lengths, minimum stomatal resistances tend to be larger than for mesophytic plants.

For example, in one study by Ting et al. (1967) with *Kalanchoe blossfeldiana*, under a variety of temperature conditions, minimum r_s estimates ranged from 2 to 10 s cm^{-1}. Studies by Szarek (Szarek et al., 1973; Szarek and Ting, 1974b) and more recently by Hanscom with *Opuntia basilaris* have shown r_s minima rarely below 2 s cm^{-1}, and on the average 5–10 s cm^{-1}. For pineapple, minima are 7–10 s cm^{-1} (Aubert, 1970, 1971; Bartholomew, 1975). In all cases, r_s can exceed 100 s cm^{-1} and may be as much as several hundred (Neales et al., 1968).

If we assume a minimum stomatal resistance of 5.0 s cm^{-1} and an r_a of 1.5 s cm^{-1} for a typical succulent plant in comparison with a mesophyte with a minimum r_s of 0.5 and an r_a of 1.0 s cm^{-1}, and calculate transpiration as a function of stomatal opening, it is noted that the result of high minimal r_s is a greater degree of stomatal control of transpiration (Fig. 5.23). Changes in stomatal opening and hence, r_s are accompanied by corresponding changes in water loss. When, however, r_s minima are low and close to r_a, large stomatal apertures lack such control.

5.3.3.4 Cuticular Resistance (r_c)

Important in succulents is also the resistance of the cuticle to gas transfer, r_c. The latter is a parallel resistance with r_s.

There have been few measurements of cuticular resistance to gas transfer in succulent plants. By estimating water loss for cacti with stomata closed, Johnson estimated 619, 624, and 1022 s cm^{-1} for *Opuntia basilaris*, *O. acanthocarpa*, and *O. bigelovii* respectively (Ting and Szarek, 1975). These extremely high resistances to water loss must not only be due to the hermetically sealed stomata, but also to the heavy deposition of wax on the stoma-bearing surface (Fig. 5.24).

Fig. 5.24. Scanning electron micrographs of the surface of *Mammilaria (upper)* and *Ferocactus (lower)* showing stomata and waxy deposition on surface

5.3.4 Response of Stomatal Movements to the Age of the Plant and Environmental Factors

5.3.4.1 Age

Kalanchoe blossfeldiana grown in growth chambers on 8-h days (light $= 5 \cdot 10^4$ ergs cm^{-2} s^{-1}) and 26° C day, 21° C night with a relative humidity of 50% showed more stomatal opening during the day than night. Leaf-resistance determinations from leaves of three different ages showed 12.3 ± 0.6, 15 ± 2.5, and

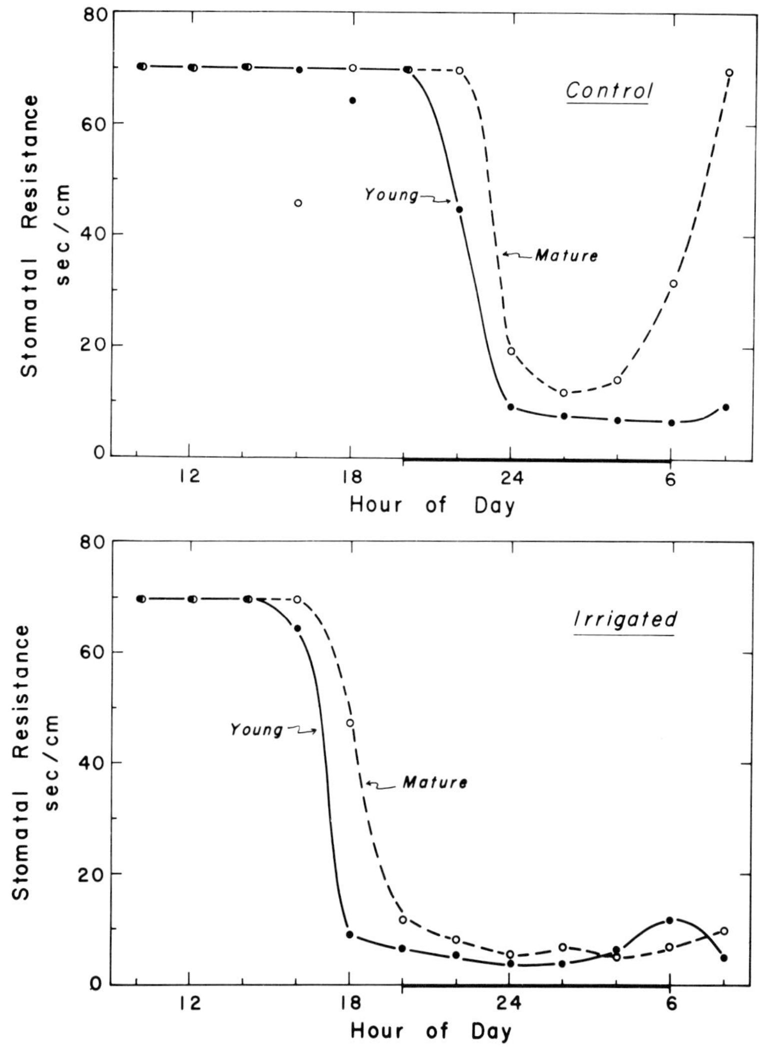

Fig. 5.25. Comparison of r_s for young and mature stems of *Opuntia basilaris* plants growing at the Boyd Desert Research Laboratory. In both irrigated and nonirrigated plants, the stomatal resistance of young tissue is less than for older, mature tissue. Data obtained by Hanscom with a porometer (Hanscom and Ting, 1977, by permission)

99.1 ± 8.8 s cm^{-1} for young, medium, and old leaves respectively. Leaf resistance was generally lower on the upper surface than the lower despite fewer stomata, 3274 ± 108 compared to 4698 ± 118 per cm^{-2} (Ting et al., 1967).

Hanscom and Ting (1977) studying *Opuntia basilaris*, found a generally lower r_s for young stem sections compared to old, regardless of environment (Fig. 5.25).

Jones (1975) using *Bryophyllum fedtschenkoi* also found lower resistances with young leaves. The young leaves which do not show CAM as evidenced by a lack of organic acid fluctuation and dark CO_2 uptake, showed low stomatal resistance

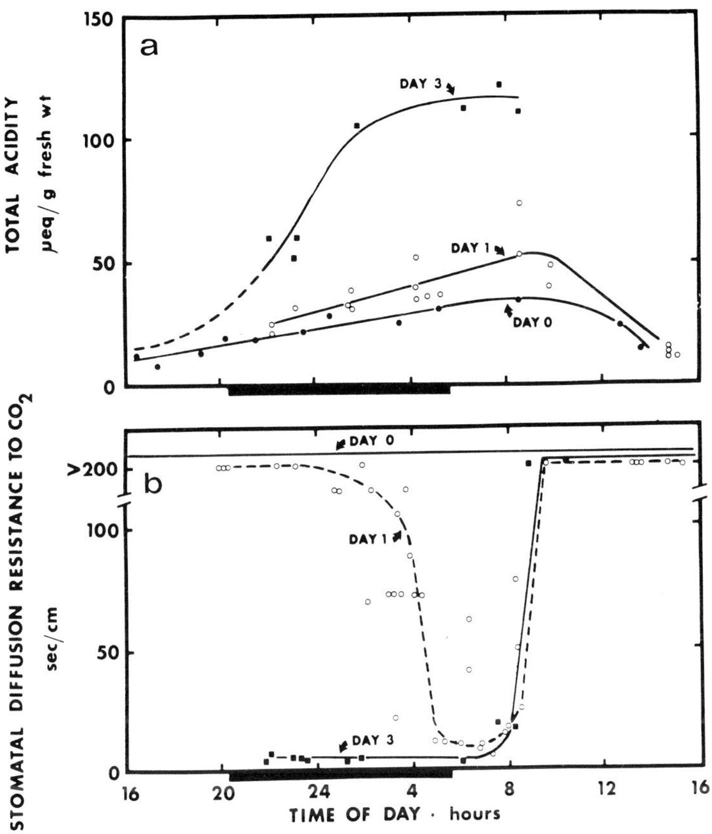

Fig. 5.26. Stomatal response to precipitation by *Opuntia basilaris*. After rains, stomata open **b**, fixation of exogenous CO_2 begins, and organic acid fluctuation **a** increases (from Szarek et al., 1973, by permission)

during the day and high at night. With the older mature leaves, the pattern was reversed. Stomata were open at night, as judged from r_s estimates, and closed during the day. The latter was accompanied by dark CO_2 uptake and organic acid fluctuation.

5.3.4.2 Water Relations

In many succulent plants, particularly within the Cactaceae, stomatal opening is a function of water status. In a comprehensive study, Szarek studied stomatal resistance to water vapor transfer in cacti at the Boyd Desert Research Center, Palm Desert, California. His data indicate (Szarek et al., 1973; Szarek and Ting, 1974b, 1975a, b; Ting, 1976b) that stomata open in response to precipitation. Following rain, ephemeral roots reform and physiological functions begin (Kausch, 1965). In *Opuntia basilaris*, as soil rehydrates, plant water potentials increase to near −2 to −3 bars, and minimum stomatal resistances decrease to 2–

5 s cm^{-1} from over 200 s cm^{-1} (Szarek et al., 1973). As shown in Figure 5.26 (cf. also Fig. 6.7), the decreased stomatal resistance is accompanied by an increased diurnal acid fluctuation.

A second study by Szarek (Szarek and Ting, 1974a) confirmed that the stomatal response to precipitation is consistent throughout the year, and relatively independent of thermoperiod and photoperiod (Fig. 6.7).

Conde and Kramer (1975) followed stomatal resistance in *Opuntia compressa* as a function of vapor pressure deficit (VPD) of the air and found that at high VPD stomata closed and at low VPD, stomata opened. When the VPD was high in the light and low in the dark, day stomatal resistance was greater than night stomatal resistance. If in the daylight period VPD was reduced, then day stomatal opening occurred. These workers suggested that the observations of James (1958) in Great Britain and Holdsworth (1971) in New Zealand, that *Opuntia* stomata were open during the day and not at night, could be due to the low VPD of these temperate climates.

Also the stomata of *Tillandsia recurvata* sensitively and directly respond to changes in air humidity during the phase of nocturnal CO_2 fixation (Lange and Medina, personal communication).

Relative humidity effects on stomatal aperture and stomatal resistance are not without precedent (Lange et al., 1971; Hall and Kaufmann, 1975), and are consistent with the data for succulent plants. This response to atmospheric humidity may account for our failure to observe significant day opening of *Opuntia* stomata even under optimum soil and plant water status. In the deserts of the southwestern USA, stomata may open during the early part of the light period, but close by midday (Szarek and Ting, 1975a, b). By and large the opening here is a response to increased plant water potential. Perhaps if atmospheric humidity were to stay high, cactus stomata would stay open throughout the day.

5.3.4.3 Temperature

In many CAM succulents, temperature influences the degree of organic acid accumulation. Nishida (1963) reported that temperature influenced succulent plant stomata in a similar manner. With *Bryophyllum calycinum*, nocturnal stomatal opening was greatest at low temperatures (12° C day/night). With increasing temperature to 27° C, there was a reduced nocturnal opening. At 22° and 27° C, some midday opening was apparent.

Agave americana responds to high night temperature by nocturnal stomatal closing and some day opening, whereas the usual pattern with cool nights and warm days is night opening and day closing (see Fig. 5.9; Neales, 1973a). At constant dark temperatures of 15° C, increasing day temperatures from 15° to 27° C resulted in day stomatal closing of *Kalanchoe blossfeldiana* stomates (Fig. 5.27; Ting et al., 1967).

Hence, both day and night temperatures influence stomatal resistance in succulents. In all experiments that have been reported, however, it is difficult to sort the effect of temperature from that of water status and mesophyll carbon metabolism influenced by temperature.

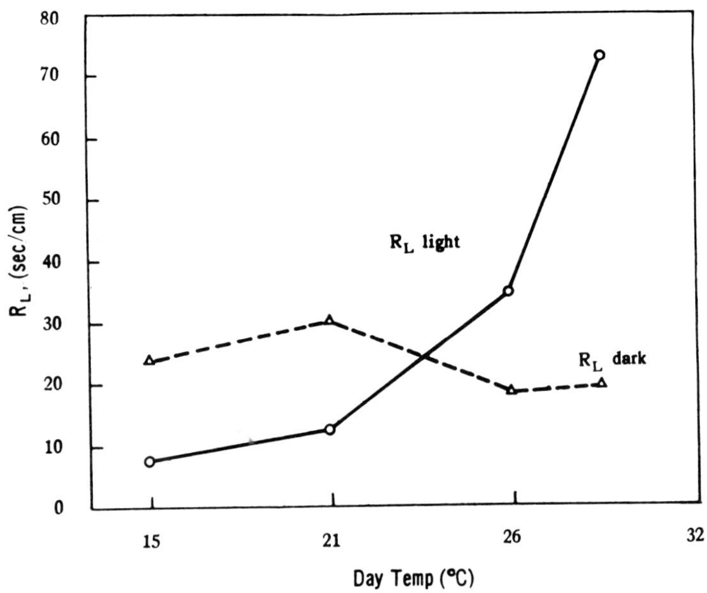

Fig. 5.27. Response of *Kalanchoe blossfeldiana* stomata to increasing day temperatures at constant night temperature. Night temperature was 15° C. Minimum night stomatal resistance, measured at the beginning of the night period, remained fairly constant, whereas day stomatal resistance increased with increasing day temperature, resulting in day stomatal closure relative to night at high day temperatures (redrawn from Ting et al., Amer. J. Bot. *54*, 245–251, 1967, by permission)

5.3.4.4 Light

In Nishida's 1963 study, he observed that a prelight period was necessary for nocturnal stomatal opening (similar to organic acid accumulation) and that the length of time stomata were open at night, as well as the maximum aperture, were functions of the duration of light in its previous light period. At high light (30,-000 lux), stomata opened in the day as well as night.

Although perhaps not relevant, Kuiper (1964) studied the wave length dependence of light on day stomatal opening maintenance in the succulent plant, *Senecio odoris*. He found that wave lengths of 432 and 675 nm were most effective, leading him to conclude that photosynthesis maintained opening.

Queiroz's research indicates that with *Kalanchoe blossfeldiana* var. Tom Thumb, short photoperiods induce nocturnal stomatal opening (Queiroz, 1974). In fact, short photoperiods induce the entire spectrum of CAM in Tom Thumb, including leaf succulence, diurnal organic acid fluctuation, and induction of CAM enzyme activity. Cause and effect relationships are not clear, but in *K. blossfeldiana* (not Tom Thumb), red light interruption of long-night plants did not alter the stomatal opening pattern of plants grown in day 26°/night 21° C thermal regimes. Stomata were open during the day and closed at night.

Table 5.4. Number of stomata per cm^2 on the transpiring surfaces of a few selected succulents[a]

Species	Abaxial	Adaxial
Agave americana	2,100	2,100
A. desertii	2,000	1,800
Aloe arborescens	2,400	3,500
Bromelia sp.	1,000	1,500
Crassula argentea	3,300	2,400
Cymbidium sparkesii	1,500	1,100
Dudleya lanceolata	1,600	2,000
Euphorbia sp.[b]	2,100	—
Hymenocyclus sp.	2,400	3,000
Kalanchoe blossfeldiana	4,700	3,300
Kleinia repens	3,200	3,200
Opuntia acanthocarpa[b]	1,500	—
O. bigelovii[b]	6,500	—
O. littoralis[b]	3,700	—
Portulacaria afra	3,300	4,500

[a] The mean estimate for the 15 randomly selected CAM succulents was $2680 \pm cm^{-2}$. These data are considered representative for succulents and account in part for the generally higher stomatal resistance.
[b] Estimates from stem tissue.

5.3.5 Morphology of Stomata in CAM Plants

5.3.5.1 Number and Distribution of Stomata

By and large, succulent plants have fewer stomata distributed over exposed surfaces than most mesophytic, nonsucculent plants (Table 5.3). Verduin (1949) estimated an average stomatal density of about 10,000 cm^{-2}. Our specific estimate for *Kalanchoe blossfeldiana* was 3274 ± 108 cm^{-2} for the adaxial surface and 4698 ± 118 cm^{-2} for the abaxial surface (Ting et al., 1967; see Table 5.4).

5.3.5.2 Size and Shape of the Stomata

The general dimensions of the stomata of succulent plants do not appear significantly different from mesic plants. Most of those that we have observed are either only slightly or not at all sunken, but are associated with a deep substomatal cavity because of the presence of a multilayered hypodermis below the epidermis. To what extent this cavity lengthens the diffusion path length and increases r_s is not known.

In *Yucca* and *Nolina* at least, the stomata are sunken within crypts which can open and close with finger-like projections (Fig. 5.28). During water stress, the leaves curl and seal the stomata within the crypts.

In many of the cacti, scanning electron micrographs reveal unsunken surface stomata not too unlike those of most mesic plants.

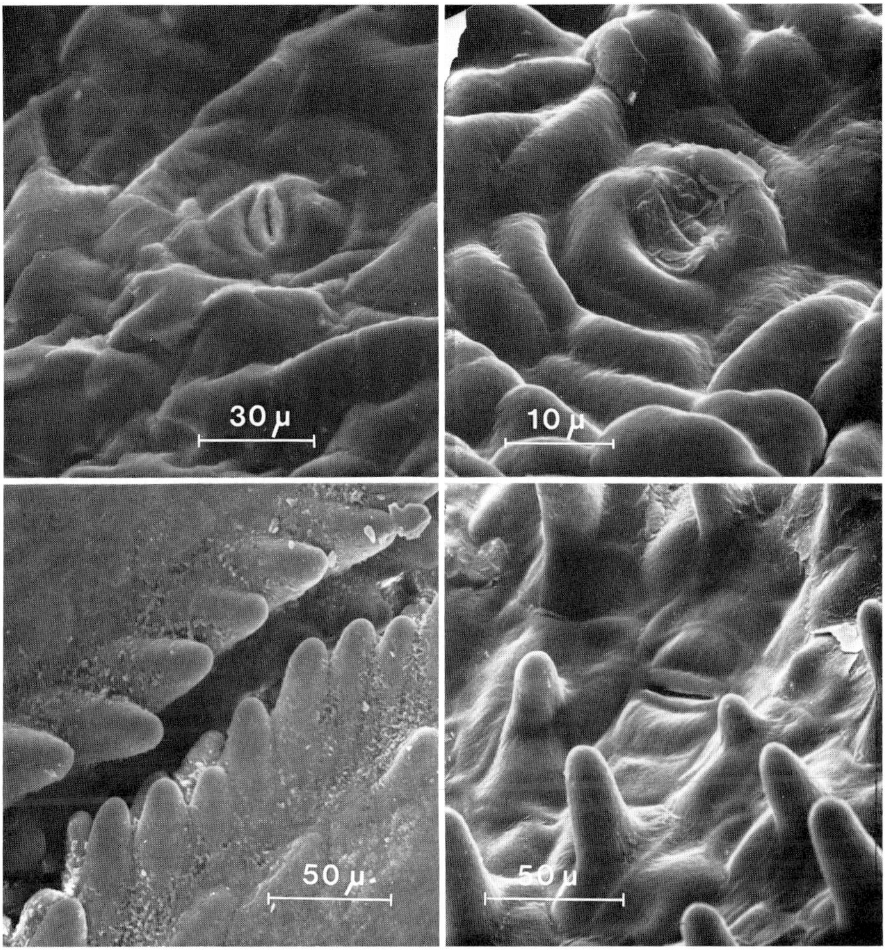

Fig. 5.28. Scanning electron micrographs of the surfaces from succulent plants. *upper left*: *Mammilaria tetrancistra*; *upper right*: *Opuntia acanthocarpa*; *lower left*: *Nolina parryi*. The stomata are located within the crypts; *lower right*: *Opuntia basilaris*

Thomson and de Journett (1970a, b) studied the guard cells of *Opuntia ficus-indica* with a transmission electron microscope (Fig. 5.29). They noted that the stomata were slightly sunken below the surface. The guard cells were suspended over large substomatal cavities by attachment to prominent subsidiary cells. They did not observe plasmodesmata connecting guard cells and subsidiary cells, nor were there ectodesmata visible.

As can be seen from the micrograph of Figure 5.29, the guard cells have large, central nuclei, plastids with starch grains and numerous mitochondria in a ratio of about eight mitochondria to each plastid. Endoplasmic reticula, microbodies, and numerous vacuoles are present. Thus, these guard cells and stomata of *Opuntia* are similar to those of other plants.

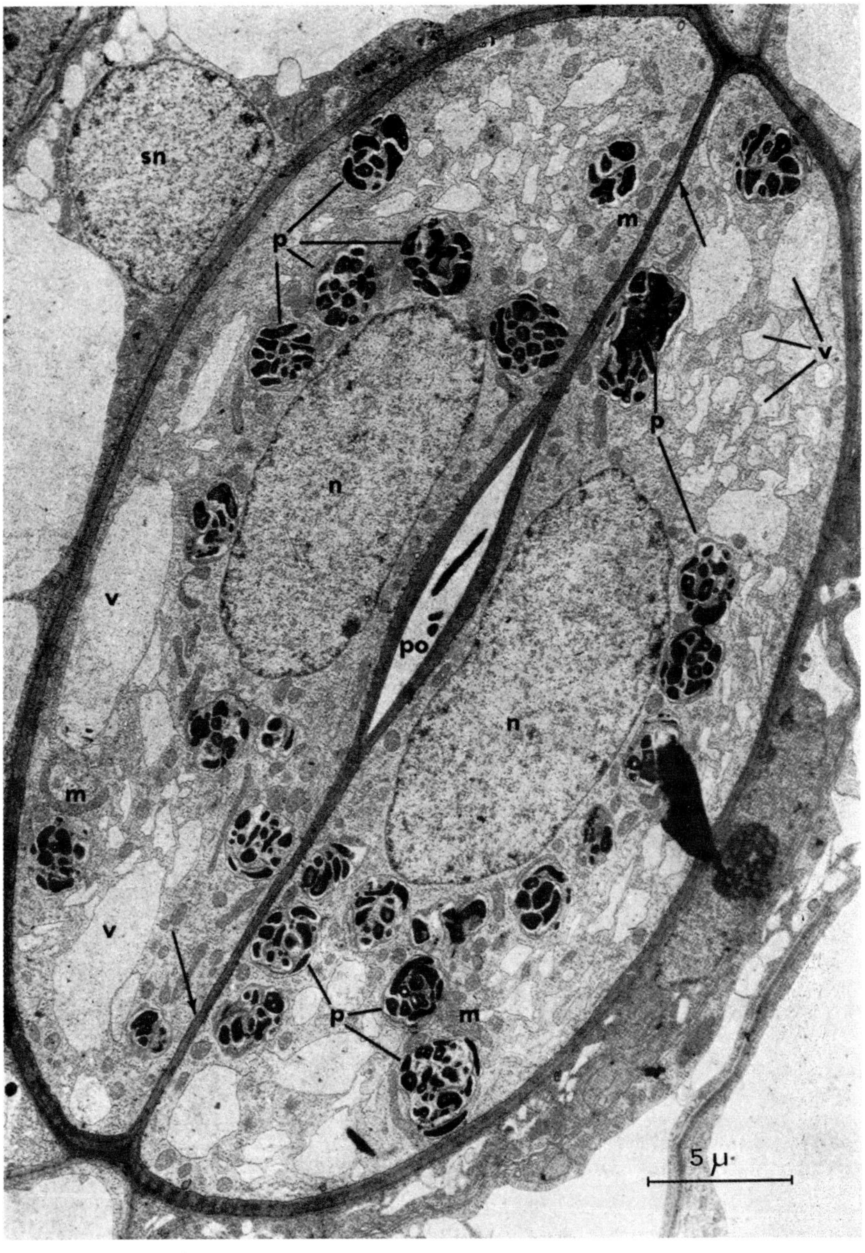

Fig. 5.29. Transmission electron micrograph of a stoma on stem of *Opuntia ficus-indica*. *V*, vacuoles; *m*, mitochondria; *n*, nucleus; *p*, plastids; *po*, stomatal pore (slightly open); *sn*, nucleus of a subsidiary cell (from Thomson and De Journett, 1970a, by permission)

5.3.6 Thermal Consequences of Stomatal Behavior in CAM Plants

Gates (1962) has developed and expounded on the thermal balance of plant tissues in general. The heat budget of tissue can be described according to Gates and Benedict (1963) as:

$$Q_A = Q_R + Q_C + Q_T \quad (5.3)$$

where
Q_A = total energy absorbed
Q_R = energy lost by reradiation
Q_C = energy lost by convection
Q_T = energy lost by evaporation

The magnitude of the Q_T term is a function of Eq. (5.1) and the magnitude of Q_C is a function of the tissue/air temperature differential and the resistance of the boundary layer to heat transfer. Hence, the succulent morphology tending toward a spherical shape and consequent increase in boundary layer resistance tends to increase the thermal load of the tissue, since the boundary layer would tend to decrease both convective (Q_C) and evaporative (Q_T) heat losses. Further, with stomata closed during the day, greatly impending transpiration, further reduction in evaporative cooling is likely. It is not uncommon therefore for succulent tissue to be 20° C or more (Huber, 1935) above air temperature when in direct sunlight (Larcher, 1973, Gibbs and Patten, 1970; Baskin and Baskin, 1971, 1973).

The Q_C term of Eq. (5.3) is to some extent a function of air movement over the surface. The fluted shape and spiny nature of some succulents, particularly cacti, would tend to reduce convective energy losses (Gibbs and Patten, 1970).

The instantaneous rate of heating (and cooling) (H) of plant tissues was investigated by Ansari and Loomis (1959). They observed that the instantaneous rate of heating was a function of the total incoming radiation (R), the specific mass of the tissue (M), and the absorptivity (A),

$$H = R/(M \cdot A)\,. \quad (5.4)$$

Thus succulent tissues with greater mass were found to heat and cool more slowly than nonsucculent plant tissues. The equilibrium temperature of the succulents, however, is greater than nonsucculents (Ansari and Loomis, 1959).

Succulents can, nevertheless, withstand great thermal stress. MacDougal (1921) reported growth of *Opuntia* between 0° and 50° C. Perhaps the high mucilage content of many succulents (Spoehr, 1919) may protect against heat denaturation of protein (Henckel, 1964). High mucilage will protect enzyme activity from heat denaturation in a manner similar to bovine serum albumin (Sutton and Ting, unpublished). Schwemmle and Lange (1959); Lange and Schwemmle (1960) found in *Kalanchoe blossfeldiana* a diurnal rhythm of heat resistance. The minimum resistance was found during the middle of the light period. Highest resistance occurred during middle of darkness. It is tempting to speculate that sensitivity against heat is highest during those times when the cytoplasm can be expected to have high concentrations of malic acid. The above resistance was generally higher in flowering (i.e., succulent) specimens than in nonflowering (i.e., less succulent) ones.

As indicated earlier, the succulent morphology tending toward a spherical shape tends to reduce evaporative water loss. This same adaptation for water conservation tends to reduce the capacity for CO_2 uptake. Further, the morphology and physiology result in higher heat loads because of greater specific mass, and reduced evaporative cooling. The in vivo thermal relations of *Capiapoa haseltoniana* (Cactaceae) growing at its natural habitat in the Atacoma desert were studied by Mooney et al. (1977b). These plants are oriented in such a manner that they reduce the high solar radiation of noon time. The epidermal surface of this cactus is highly wettable by fog and dew moisture and highly reflective when dry, thus reducing the temperature difference between plant and surrounding air. However, the meristems of this plant are located so that they have high temperatures with respect to the environment. This could increase growth rate without endangering the water use efficiency.

Note: Some recent papers dealing with the subjects of Chap. 5 are discussed in the Appendix on p. 201.

6. Ecology, Productivity, and Economic Use of CAM Plants

6.1 The Hypothesis: Ecological Advantage of CAM

As far as is known, Wilhelm Pfeffer (1845–1920) was the first to recognize that CAM might have an ecological significance. Pfeffer interpreted the nocturnal acid accumulation of succulents as a mechanism to conserve carbon in those plants which have large CO_2 diffusion distances and a relatively small number of stomata. The main sentences of Pfeffer's postulate are cited here in the original from Pfeffer's *Pflanzenphysiologie* (1897): „Für diese Pflanzen, bei denen der Gasaustausch und insbesondere die Gewinnung der Kohlensäure erschwert ist, hat es ohne Frage Vortheil, daß während der Nacht möglichst wenig Kohlensäure durch den Atmungsprozeß verloren geht. Das ist aber damit erreicht, daß zunächst organische Säuren entstehen, die am Licht allmählich weiter zu Kohlensäure oxidiert werden, welche dann ohne zuvorigen Austritt aus dem Gewebe der Verarbeitung im Chlorophyllapparat anheim fällt[3]."

Pfeffer's interpretation of CAM is an excellent example of scientific intuition. The biochemical mechanism of CAM, and in particular the occurrence of carbon dioxide fixation in darkness and the pathway of malic acid conversion to carbohydrates were unknown when Pfeffer wrote his *Pflanzenphysiologie*.

Nevertheless, some of the basic points of Pfeffer's interpretation of CAM can still be accepted. The rediscovery of dark CO_2 fixation gave no reason to change the basic concept that CAM is a mechanism to recover respiratory CO_2 during the night (cf. the reviews of Wolf, 1960; Ranson and Thomas, 1960). It will be shown later that in fact in special cases, CAM photosynthesis can function nearly exclusively on recycled respiratory CO_2 (see Chap. 6.2.2).

However, as indicated earlier, Nuernbergk (1959, 1961, 1962) was the first to recognize that the nocturnal fixation of external CO_2 may be at least as important for the final carbon balance of a CAM plant as the refixation of respiratory CO_2. It will be shown later that this proposal has been fully verified by the recent studies of CAM plants in their natural habitats.

---

[3] In those plants, where the gas exchange, in particular the gain of carbonic acid is rendered difficult, it might doubtless be advantageous that during the night the loss of carbonic acid by respiration is kept as low as possible. This, however, is achieved by first synthesis of organic acids, which in the light become oxidized step by step to carbonic acids. This carbonic acid becomes available to the chlorophyll-apparatus for further conversion without having left the tissue.

It is now generally accepted that the overall carbon balance is only one aspect of CAM ecology. As indicated in Chapter 1, CAM plants are distributed in many of the deserts of the world, especially in North and South Africa, North and South America, and West Australia. Furthermore, the epiphytic CAM plants, predominantly bromeliads and orchids, exist in locally dry environments. Others still, such as many crassulacean succulents, although perhaps existing in seemingly mesic habitats, actually occur within rocks or in shallow soils with prolonged water deficiencies. Hence, it can be predicted, that water economy is an additional important ecological aspect of CAM. In this sense, the ecological significance of the CAM photosynthetic pathway appears to be an adaptation to arid or otherwise dry environments.

The specific adaptation seems to be severalfold (Kluge and Fischer, 1967; Szarek and Ting, 1974a, 1975a, b; Kluge, 1971c, 1973, 1976, 1978; Ting et al., 1972; Ting, 1976b). Firstly, a generalization for all plants, for photosynthesis to occur stomata must be open because it is through the stomata that CO_2 enters. A great paradox of plant biology is that when stomata are open, vast quantities of water are lost. For mesic plants of the C_3-photosynthetic option type, 600–1000 times more water is frequently lost per volume of carbon dioxide assimilated. Because of an apparent greater carboxylation capacity, the C_4-photosynthetic option plants have a reduced ratio to an average of about 300 (see Black, 1973). CAM plants functioning in the CAM mode by fixing CO_2 predominately at night, when stomata are open, are necessarily under less evaporative demand than they would be if CO_2 were assimilated during the daylight hours with a consequent high evaporative demand. Then, by being tightly sealed during the daylight hours, CO_2 is assimilated with minimum water loss. This reverse phase of stomatal opening alone tends to reduce the ratio of water loss to carbon dioxide gained to an average of 10–150, at least 50% less than C_4-plants.

By way of example, assuming a CO_2 assimilation rate of 10 mg dm^{-2} h^{-1} day or night and a transpiration ratio (see Chap. 6.2.1) of 300 during the day, the transpiration rate would be 3 g dm^{-2} h^{-1}. If, as is frequently the case, the temperature of the plant during the day were 45° C, the vapor pressure would be 90 m bars. A reduction of temperature at night to 20° C would reduce the vapor pressure to 20 m bars. Hence, transpiration at night would be four to five times less than during the day, and the transpiration ratio at night would be 75.

Secondly, CAM maintains a positive carbon balance or at least prevents a negative balance even during extended periods of drought. This is evident in comparative experiments of net CO_2 exchange under drought conditions (Fig. 6.1). In such experiments, mesic plants have a negative carbon balance after relatively short drought treatments. In contrast, because of water storage in large cell vacuoles, together with CAM metabolism as a water-saving mechanism (see above), CAM plants continue to harvest CO_2 from the atmosphere for long periods even if irrigation is lacking.

Thirdly, CAM plants can respond to drought by hermetically sealing the stomata, virtually eliminating water loss. In the case of cacti at least, what little water is lost from young stems is replaced by transfer from older tissue. An experiment illustrating this phenomenon was conducted at the Boyd Desert Research Center by Stan Szarek. In June, 1971, *Opuntia bigelovii* plants with about 40 stem sections

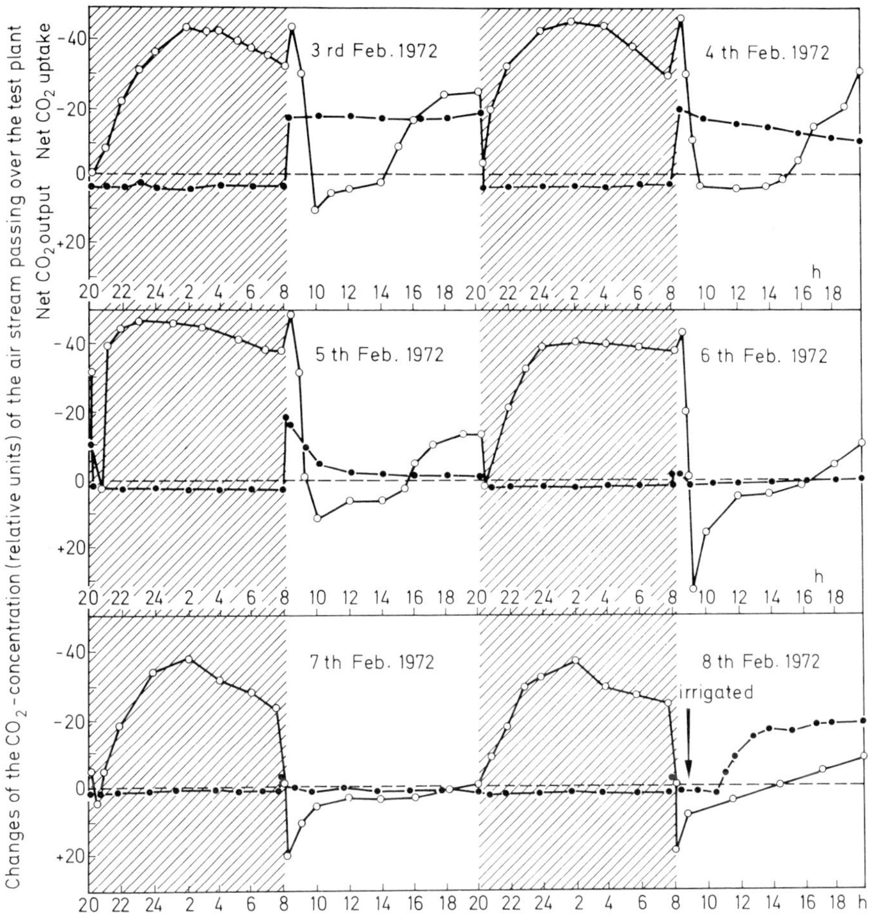

Fig. 6.1. The diurnal cycle of CO_2 gas exchange of *Kalanchoe daigremontiana* (CAM) and *Coleus* spec. (C_3) under artificial drought. ○—○ = *Kalanchoe*, ●—● = *Coleus* (from Kluge, 1976a)

were severed at the base and supported in situ on ring stands. At this time, the plants were under some water stress with stem water potentials of about −8 bars and a mean organic acid diurnal fluctuation of 20 meg per kg fresh weight. In June, 1974, the five remaining green stem sections had a water potential of −13 bars and a mean organic acid fluctuation of 15 meg per kg fresh weight (Fig. 6.2).

Transfer of water can be observed also in other succulents. Such an example has been described by Nobel (1977a). An individual of *Agave deserti*, during the 159 days between the occurrence of the flower and the ripening of the seeds, lost 24.9 kg fresh weight of the leaves. Only 7.1 kg were due to water loss by leaf transpiration. The bulk of water lost from the leaves appeared in the inflorescence where it was stored (3.1 kg), or was lost by transpiration (4.3 kg), or remained in the fruits (10.8 kg), respectively.

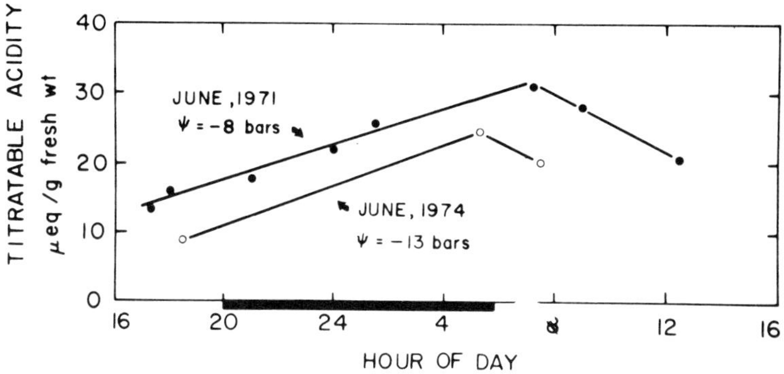

Fig. 6.2. Diurnal fluctuation of titratable acid in *Opuntia bigelovii* stems before and after 3 years since severing at the base (from Szarek and Ting, 1975a)

During extended natural periods of drought there is little or no measurable water loss or CO_2 assimilation. The plants apparently maintain a favorable carbon balance by continually recycling internal CO_2 through the CAM pathway (Szarek et al., 1973). It is a unique property of the CAM plants that they are able to persist for extended periods without net CO_2 exchange (see Fig. 6.7).

The sealing to prevent excessive water loss evidently comes about by tightly closed stomata and thick impervious cuticles presenting a barrier to water loss. Many of the succulents when older have an extensive amount of wax on their surface. These plants as they develop go through a young, very green stage to an old white scurfy form because of wax buildup (Fig. 5.24). The resistance to water vapor diffusion can be in the vicinity of 1000 s cm^{-1} (Szarek and Ting, 1974b).

Finally, as detailed elsewhere (Chaps. 4.3.4.2; 5.1.3; 6.2.2.3), in addition to CAM per se, there is a growing literature suggesting that certain CAM plants can shift from the C_3 photosynthetic option to the CAM mode, allowing maximum exploitation of a changing environment.

In the following, the ecological significance of CAM postulated above will be proved by the discussion of water relations in CAM plants and with a consideration of CAM featured by plants growing in situ in their natural environments.

6.2 Verification of the Hypothesis

6.2.1 CAM and Water Use

The effectivity of water use by plants is conveniently expressed in terms of transpiration ratio (TR), the ratio of grams water loss into grams CO_2 assimilated. Using the gas-transfer equation 5.1, transpiration can be expressed as:

$$T = \frac{\Delta}{r_a + r_s} \tag{6.1}$$

where

T = transpiration rate in g H_2O loss dm^{-2} h^{-1}

Δ = vapor pressure difference between tissue evaporating surface and air sink

r = as described for Eq. (5.1).

Similarly, CO_2 assimilation can be described:

$$P = \frac{\Delta'}{r_a + r_s + r_m} \tag{6.2}$$

where

P = g CO_2 uptake $dm^{-2}\,h^{-1}$

Δ' = CO_2 concentration difference between air source and tissue sink. Usually, it is assumed that the ambient air is constant at about 300 ppm CO_2 or 58.9 mg $CO_2\ cm^{-3} \cdot 10^{-5}$ at standard pressure, and the tissue sink is 0. Alternatively, Δ' can be calculated as internal CO_2 minus the CO_2 compensation point.

r = as above (for typical r values see Table 6.1).

The transpiration ratio is, as defined here, the ratio of T to P. A main problem in interpretation, however, is that T is tightly coupled to the environment through Δ. Hence, a graph of T versus Δ would be linear if the diffusion resistances are constant. Similarly, a graph of T:P versus Δ would be linear and hence, the transpiration ratio (T:P) will vary with the evaporative demand of the environment.

In comparing transpiration ratio, it is assumed that Δ' is constant among the test species. Actually, the slope of the graph T:P versus Δ gives a better indication of water use efficiency in terms of carbon assimilated.

The slope of the line, T:P versus Δ is $\Delta'(r_a + r_s + r_m / r_a + r_s)$ and is related to a function E which is defined as the ratio of maximum CO_2 assimilation (P_{max}) to actual CO_2 assimilation (Ting et al., 1972).

P_{max} is defined as

$$P_{max} = \frac{\Delta'}{r_a + r_s} \tag{6.3}$$

or P at minimum r_m. Those plants with high transpiration ratios are expected to have low E ratios and vice versa.

Transpiration ratios have been estimated for a variety of CAM plants. Joshi et al. (1965) studied pineapple growing in a glasshouse in North Carolina during the winter. They estimated a CO_2 uptake rate of 0.5–0.7 mg $dm^{-2}\,h^{-1}$ and transpiration of 20–50 mg $dm^{-2}\,h^{-1}$. This gives a TR range of 40–72. For an 80-day period, pineapple growing at a rate of 15–18 mg dry weight increase per day dm^{-2} leaf showed a ratio of about 50 for actual water loss to dry matter accumulated.

Neales et al. (1968) estimated Transpiration Ratios for *Aeonium hawarthii*, *Agave americana*, and pineapple in comparison with sunflower and tobacco. Over the day/night cycle under their experimental conditions, they estimated TR of 154 (*Aeonium*), 47 (*Agave*), and 116 (pineapple) for the CAM succulents and 166 (sunflower) and 162 (tobacco) for the mesophytes. Erler (1969) also reported a low TR for *Agave americana* of 70, and Sideris and Krauss (1955) estimated 55 for pineapple.

Meinzer and Rundel (1973) in greenhouse experiments with *Echeveria pumila* estimated a mean TR of 135 from the ratio of water loss to dry weight and a mean ratio of 134 from instantaneous CO_2 assimilation experiments.

Table 6.1. Estimates for r_s and r_m reported in the literature

	CAM		C_4	C_3
	Dark	Light		
r_s (minimum) (s cm^{-1})	<3.0	>6.0	>1.0	<1.0
r_m (minimum)	1–2	20–40	1.5	3.0

Studies by Szarek (1974) with *Opuntia basilaris* growing in situ in the Colorado Desert of California indicated a range of TR from 50 to over 1000. In general, however, if tissue water potentials were high in the vicinity of -2 to -3 bars, then CO_2 assimilation was significant and estimated TR values were low in the range of 50–150. It was usually at high water stress, when transpiration and CO_2 assimilation were low, that TR estimates were high. Evidently the leakage of water from the nearly sealed plants far exceeds the capacity for CO_2 uptake when metabolism is low (Szarek et al., 1973).

Nobel (1977b) found in the barrel Cactus, *Ferocactus acanthodes*, growing in natural stands, TR values of 105 during the hot months and 18–24 during the cooler season.

The significance of these generally low TR estimates for CAM plants is clear when compared with other photosynthetic groups of plants. Range values for C_3 plants are 450–950 and 250–350 for C_4 plants (Table 6.3). The range estimate for CAM plants is 50 to 125 (Black, 1973; Ting et al., 1972).

The estimate E, a measure of CO_2 assimilation efficiency relative to water loss (see above), can be determined by adjusting the measured CO_2 assimilation to the value it would be in the absence of the residual resistance, r_m. Alternatively, ignoring r_a or summing r_a+r_s, the ratio r_s to (r_s+r_m) approximates E. As r_m approaches zero, E approaches 1.

Several careful studies reported in the literature allow calculation of E from resistance ratios. For example, Bierhuizen and Slatyer's (1964) data give an E value for cotton, a C_3 plant, of about 0.22. Ludlow (1970) and Ludlow and Wilson's (1972) data give 0.32 and 0.3 for the C_3 plants *Phaseolus atropurpureus* and *Vigna luteola*. In a comparative study of C_3 *Atriplex hastata* and C_4 *A. spongiosa*, Slatyer's data give 0.3 and 0.79, respectively. *Saccharum*, a C_4 species, calculates at about 0.83 (Bull, 1969), *Zea mays* (C_4) at 0.65 (from data in Gifford, 1974), and the C_4 *Pennisetum purpureum* had an E estimate of 0.72 (data of Ludlow and Wilson cited in Gifford, 1974). Björkman et al. (1972) reported data indicating a near-maximum E for *Tidestromia oblongifolia*.

Few data are available for CAM plants, however, the study of Neales et al. (1968) with pineapple showed E values of 0.72 and 0.93 for dark and light.

Estimates of E for *Opuntia basilaris* growing in situ in the desert ranged from 0.2 to 0.82. Similarly to TR, during periods of minimal water loss and reduced metabolic activity, E estimates were low; but in the dark under favorable water balance, maximum E values are high in the range of 0.8–0.9. In the light, however, E estimates are lowered to about 0.2–0.3.

Hence, the data taken together indicate that in terms of water-use efficiency CAM plants are similar to C_4 plants in that TR ratios are low. These data support the theory that CAM plants are well adapted to survival in arid lands.

The study of Meinzer and Rundel (1973) points to the importance of TR for CAM plants. Their data show an inverse relationship between growth of *Echeveria pumila* and TR (Fig. 6.3).

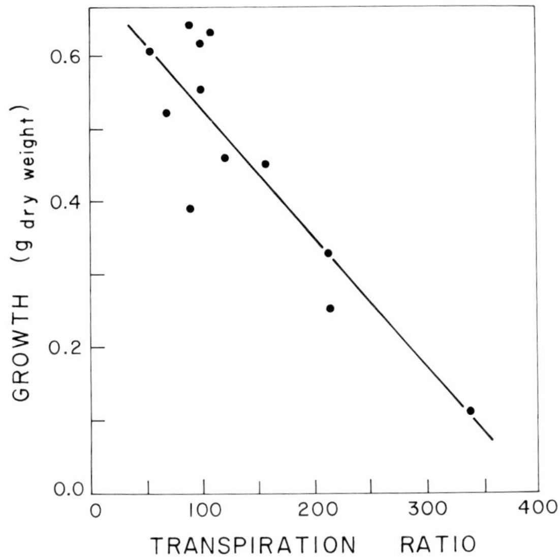

Fig. 6.3. Growth of *Echeveria pumila* correlated with transpiration ratio. Greenhouse-grown plants (from Meinzer and Rundel, 1973, by permission)

CAM plants, especially the cacti, respond rapidly to water availability. In one experiment, irrigation of cacti *(O. basilaris)* resulted in a water potential increase from about −15 to −5 bars in a little over 12 h (Fig. 6.4). These same plants respond to water by a relaxation of r_s and consequent CO_2 assimilation resulting in a marked increase in diurnal acid production (Fig. 6.6).

A study by Kausch (1965) with *Opuntia puberula* which had been dry for many months showed a rapid response to water. Absorbing roots were formed in a few hours, transpiration increased, and both dark and light CO_2 assimilation were stimulated.

In *Opuntia*, there is a direct correlation between ψ and acidification rate (Fig. 6.5).

In analogy to the $\delta^{13}C$ values which are an index of carbon isotope composition in plants (see Chap. 3.3), the hydrogen isotope composition of plant material can be characterized by a term δD which expresses the ratio of deuterium to hydrogen as follows (see Ziegler et al., 1976)

$$\delta D = \left[\frac{\text{D/H sample}}{\text{D/H standard}} - 1\right] \times 10^3 \,[‰] .$$

Ziegler et al. (1976) found that δD values of the metabolic hydrogen in CAM plants range from −75‰ to +5‰, which is less negative than in C_3 or C_4 plants. When CAM plants are treated experimentally with water stress, enrichment in deuterium occurs. The above authors assume that these changes of hydrogen isotope composition may reflect deuterium enrichment of tissue water during transpiration. They further suggest that "whatever the origin of the variable δD value in CAM plants, this parameter may be a useful index of the water relations of these plants under natural conditions." Further investigations in this field are desired.

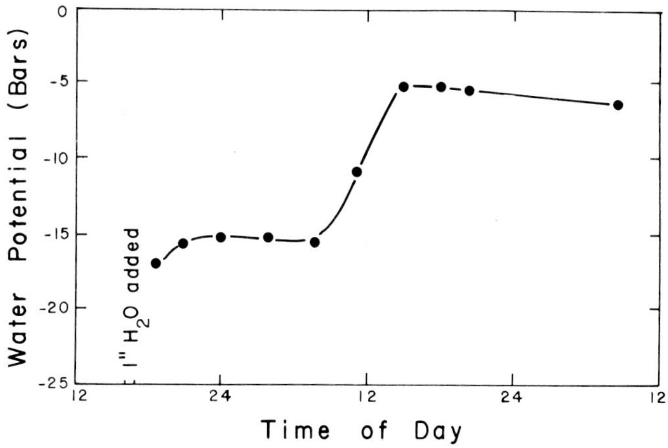

Fig. 6.4. Hydratation of *O. basilaris* after irrigation (US/IBP of Szarek, Brum, Johnson, and Ting)

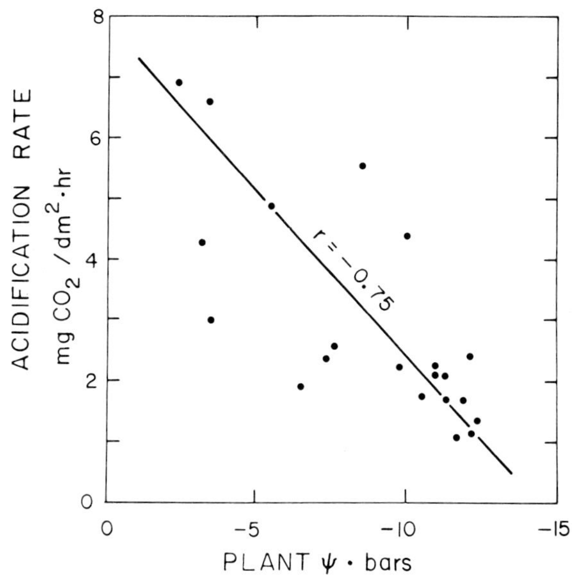

Fig. 6.5. Relationship between acidification and water potential (ψ) of *O. basilaris* (from Szarek and Ting, 1974b)

6.2.2 Observation of CAM in Situ

The question of CAM plant physiology in situ in natural habitats has recently been studied and increasingly attracts attention. Typical features of CAM plants, such as gas exchange, acid fluctuation, and stomatal behavior, have been measured continuously in the field during day/night cycles and seasonally. Also, ^{13}C isotope composition ($\delta^{13}C$ value) (see Chap. 3.3) has been determined in samples

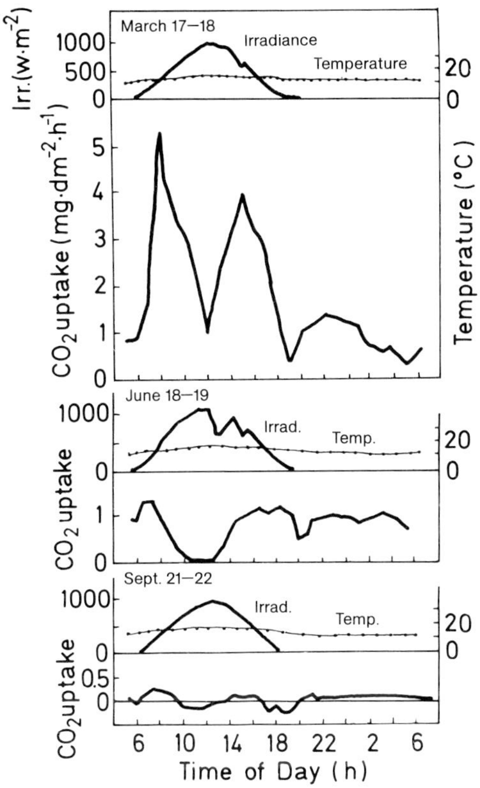

Fig. 6.6. CO_2 exchange of *Dudleya farinosa* at its natural habitat (from Bartholomew, 1973, by permission)

removed from CAM plants growing in natural habitats. As outlined earlier (see Chap. 3.3), it can be determined from ^{13}C isotope composition estimates if there is a shift from C_3 toward the CAM option of photosynthesis in response to environment. It can be predicted that the ^{13}C isotope composition will shift from C_3 to CAM as the natural plant distribution shifts from mesic to arid. Results from both physiological studies in situ and δC_{13} analysis from naturally growing plants are discussed below.

6.2.2.1 Gas Exchange and Acid Fluctuation

The first continuous measurements of gas exchange of a CAM plant in situ were provided by Bartholomew (1973). He studied *Dudleya farinosa* (Crassulaceae) growing in coastal California. Basically, the CO_2 exchange followed the typical gas exchange pattern of CAM plants known from laboratory experiments with net CO_2 uptake during the night and a depression of CO_2 during the day (Fig. 6.6). There was a clear shift in the CO_2 exchange pattern seasonally. At the beginning of the dry months, CO_2 fixation during the day exceeded nocturnal CO_2 fixation. However, the daytime CO_2 fixation was diminished during the drought period. Because the dark CO_2 fixation remained nearly unaffected under drought conditions, *D. farinosa* assimilated CO_2 from the atmosphere mainly at night, consistent with the results obtained by treating *Kalanchoe daigremontiana* with water stress (Kluge and Fischer, 1967; see Fig. 6.1), and with observations in *Echeveria* (Medina and Delgado, 1976).

The air and leaf temperature of *Dudleya* is remarkably constant during the day/night cycle and during the seasons because of the vicinity of the ocean. In spite of this, CAM is very pronounced. Thus, at least in *Dudleya*, CAM is not ultimately controlled by temperature.

When studying gas exchange of *Agave deserti* under field conditions in the Colorado Desert, Nobel (1976) showed that during the winter *Agave* had the CAM CO_2 exchange pattern and the inverse rhythm of stomata of CAM plants, i.e., there was high stomatal resistance during the day and low resistance during the night. The resistance for water vapor diffusion was correlated with the soil water potential. During the drought season, no natural stomatal opening occurred, but it could be easily induced by artificial irrigation of the plant. Nobel reported that in consequence of a shallow root system, *A. deserti* responds quickly to the infrequent rains, and because of the water-storing capacity of the succulent leaves, the CAM-type pattern of stomatal opening was maintained for about 8 days after the water potential of the soil became more negative than that of the plant.

The transpiration ratio of *Agave deserti* was estimated to be as low as 18 for winter days, and 25 as an average for the entire year. In contrast to *Dudleya farinosa* (Bartholomew, 1973, see above), temperature seemed to be important in modulating the gas exchange pattern of *Agave deserti* in the field. For instance, the stomatal resistance for water vapor diffusion increased about five fold if the leaf temperature was artificially increased from 5° to 20° C. This suggests clearly the importance of cool nights for CAM of *Agave deserti*.

Nobel's observations of *A. deserti* are in close agreement with the results obtained by Szarek et al. (1973), Ting and Szarek (1975), and Hanscom and Ting (1977), who provided observations of CAM behavior in *Opuntia basilaris* in its natural habitat at the Boyd Deep Canyon in California. Gas exchange parameters of CAM and acid oscillation were estimated (Fig. 6.7). Similarly to *A. deserti*, the plant water potential was of prime importance for CAM metabolism in the field.

By means of rapid development of shallow roots, similar to *Agave deserti*, *O. basilaris* responds quickly even to a single period of precipitation. For example, the plant water potential (ψ) followed the upper soil layers until soil ψ decreased below -40 bars. After rainfall as low as 6 mm, the plant ψ recovered to values of -1 to -3 bars (cf. Fig. 6.7, August). When the plant ψ was high, nocturnal CO_2 fixation reached values of 10–15 mg CO_2 dm^2 h^{-1}, and low stomatal resistances (2–3 s cm^{-1}) were observed during the night. Under these conditions, the day/night difference reached 100–150% and the maximum activity levels during the night approached 75–90 μeq/g fr. wt.

The transpiration ratio of *O. basilaris* varied significantly during the year (see Fig. 6.7), although mean TR values remained within the values reported for the three optional types of photosynthesizing plants, i.e., 450–600 for C_3-plants, 250–350 for C_4-plants, and 25–150 for CAM plants.

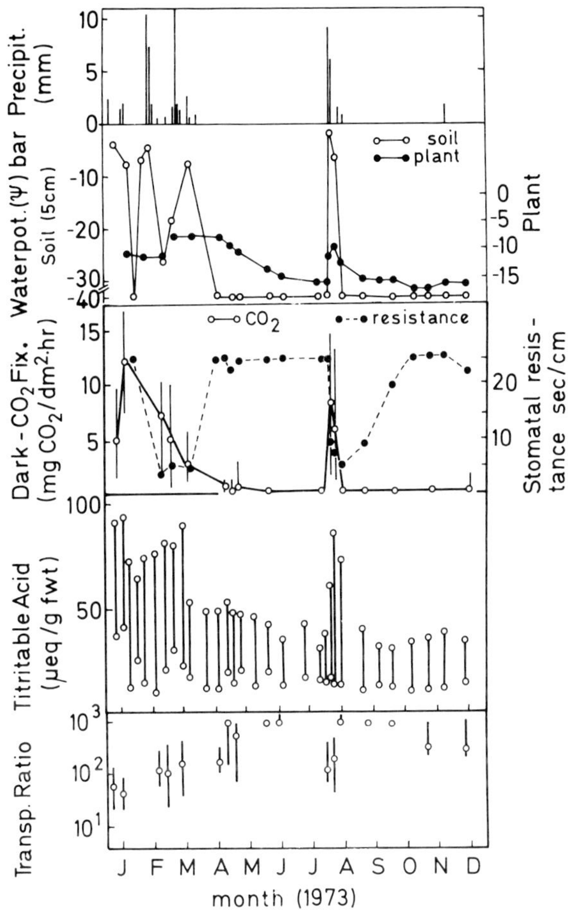

Fig. 6.7. Performance of CAM by *O. basilaris* in its natural habitat throughout 1973 (data from Szarek, 1974)

Figure 6.7 shows that during periods of rainfall and high plant ψ, the water use efficiency was highest. However, during the drought season, the TR of *O. basilaris* was above the level of C_3 and C_4 plants.

This latter behavior appears to be a paradox. However, it should be noted that during the drought period, the stomatal resistance is very high, and gas exchange extremely low. Thus, a low TR occurring during this time is ecologically insignificant.

Because of a linear relationship between plant water potential and the acidification rate (Szarek and Ting, 1974; see Fig. 6.5), the highest amplitude of acid rhythm in *O. basilaris* occurred when the plant ψ was high, and the lowest amplitudes could be observed when plant ψ were low. Nevertheless, the nocturnal acidification persists throughout the year, even during the drought period when the stomata are sealed.

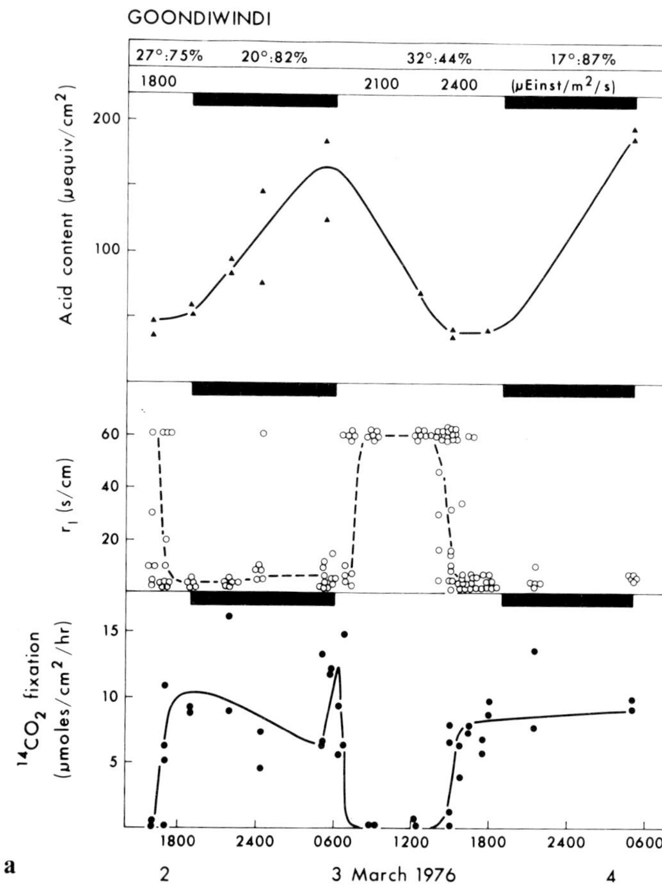

a

Fig. 6.8 a and b. Performance of CAM in *O. inermis* on the natural stands at Goondiwini and Denman (Eastern Australia). **a** Diurnal cycle of CAM (malic acid rhythm, leaf resistance, and $^{14}CO_2$ fixation). Actual temperatures and relative air humidities are indicated at top of **a**. **b** Seasonal changes in the pattern of dark CO_2 fixation and course of leaf resistance. Malic acid levels, soil water capacity (SWC) and plant water potential (ψp) are indicated in the upper parts of the figure (after Osmond et al., 1978, by permission)

It is reasonable to conclude that this persistence of CAM activity during the drought period is the result of day/night recycling of endogenous CO_2. Here, malic acid synthesis depends exclusively on CO_2 produced in respiration, and photosynthesis uses only endogenously produced CO_2 from malate decarboxylation and photorespiration.

The Barrel Cactus, *Ferocactus acanthodes*, in natural stands in the Colorado Desert, is similar to *Agave americana* and *Opuntia basilaris* (Nobel, 1977b). Water storage in its stem permitted to keep open stomata and CO_2 uptake at night for about 1½ months after the soil water potential had become lower than that of the plant, i.e., the plant had become unable to take up water from the soil. After 7 months of drought the stomata remained closed throughout the day/night cycle.

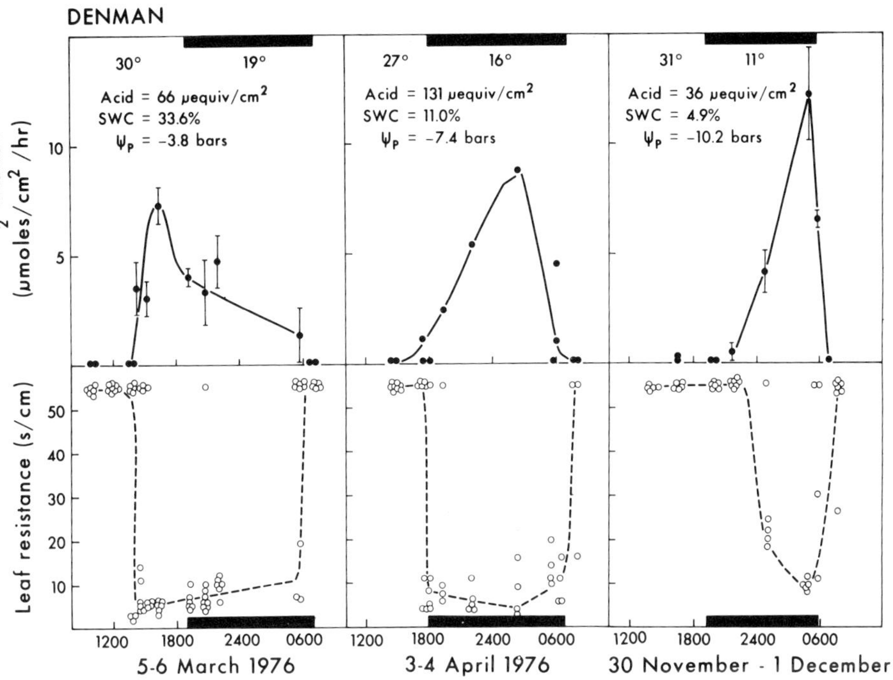

Fig. 6.8b. (Legend see opposite page)

Similarly to *Opuntia basilaris*, *F. acanthodes* has shallow roots and its water potential recovers within 24 h after rainfall. The optimal temperature for dark CO_2 fixation was 12.6° C, remarkably low. The transpiration ratio was about 70 for an entire year, which is in good agreement with values of other CAM plants (see Chap. 6.2.1). As indicated earlier (Chap. 6.2.1), during the dry season the TR values observed in *F. acanthodes* range from 18 to 105.

Osmond (yet unpublished) studied CAM in *Opuntia inermis* at various sites in Eastern Australia. Typical CAM patterns of CO_2 exchange and stomatal resistance were observed also in this species in natural stands (Fig. 6.8). These patterns undergo clear seasonal modifications (Fig. 6.8 b). During March (Australian fall), substantial CO_2 was fixed in the light, and CO_2 fixation was highest at the beginning of the night. During November (Australian summer), stomatal resistance remained high throughout the day and during the first half of the night. Only later in the night did stomata open, and net dark CO_2 fixation occur. The amounts of malic acid accumulated were correlated with the intensity of dark CO_2 fixation during the seasons.

The above examples of CAM performance under field conditions came from species growing in arid habitats in California and Australia. In the deserts of the New World, CAM plants are abundant and frequently the dominant species present. In certain other deserts, however, where precipitation is also low and irregular, CAM plants are considerably less abundant and restricted to specific habitats. *Caralluma negevensis*, such an example, was studied by Lange et al. (1975) in its natural habitat of the Negev desert (Israel).

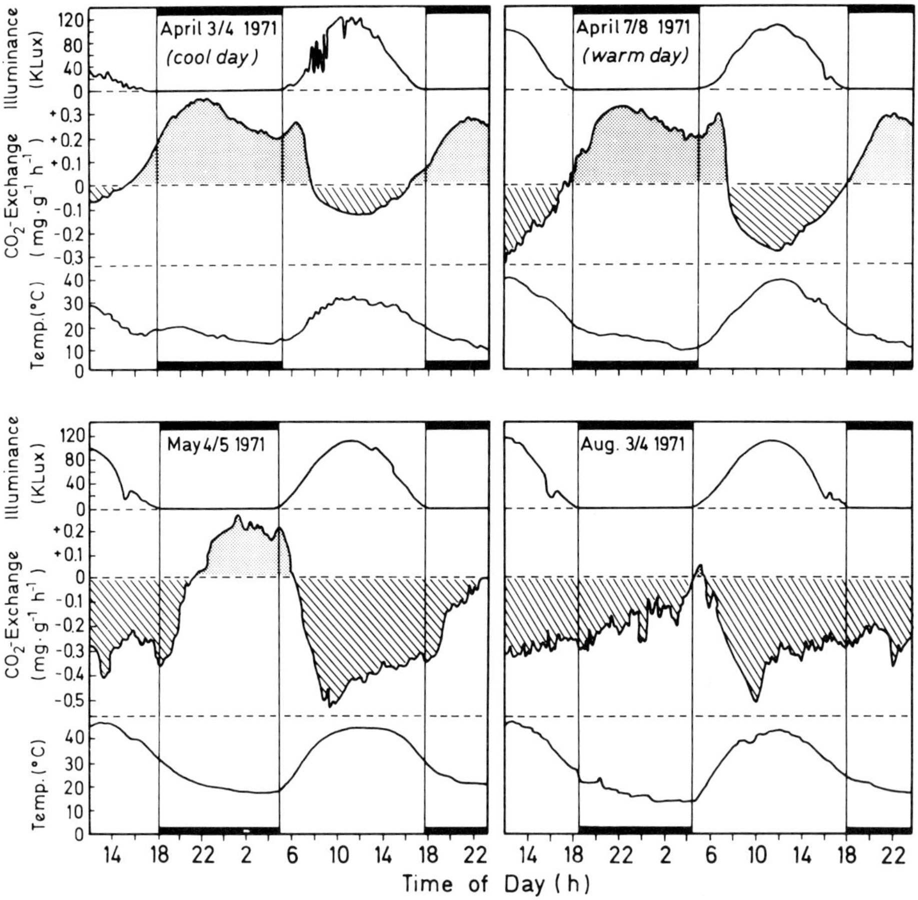

Fig. 6.9. Performance of CAM in *Caralluma negevensis* in the Negev (after Lange et al., 1975)

Table 6.2. Daily net yield (+) or loss (−) of CO_2 of stems of *Caralluma negevensis* in comparison with twigs of arido-active desert shrubs during three different periods of the season (Avdat/Negev Desert, 1967). Figures in parantheses indicate the number of days which are taken for average[a]

	CO_2 yield or loss (mg g^{-1} dry wt. d^{-1})		
	April 1–14	May 1–14	August 1–14
Caralluma negevensis	+ 1.80 (5)	− 1.92 (1)	− 7.51 (5)
Artemisia herba-alba[b]	+ 9.78 (3)	+57.61 (3)	−10.42 (4)
Zygophyllum dumosum	+23.38 (8)	+27.17 (8)	−13.87 (5)
Reaumuria negevensis	+38.74 (8)	+21.17 (1)	− 3.40 (2)
Hammada scoparia	+37.23	+65.28 (8)	+27.93 (5)

[a] Data from Lang et al., 1975.

[b] Since no other data are available, values are indicated for March 24–30 instead of April 1–14.

Fig. 6.10. *Caralluma negevensis* in its natural habitat (from Winter and Troughton, 1978a, by permission)

In April and May, the plant had high rates of temperature-dependent dark CO_2 fixation (Fig. 6.9). During most of the day, substantial net CO_2 output occurred, which was increased by temperature. In spite of the considerable net daytime carbon loss in April (last month of the rainy season), the daily carbon balance remained positive. However, beginning with May and lasting throughout the dry season, high night temperatures inhibited dark CO_2 fixation and high temperatures during the day enhanced net carbon loss by net CO_2 output. Hence, the daily carbon balance became negative. In the extreme case (middle of August), any nocturnal net CO_2 uptake was prevented, and only during the morning was there a short period of carbon gain (see Fig. 6.9).

According to Lange et al. (1975), as far as carbon gain is concerned, *C. negevensis* is less adapted to deserts than arid desert shrubs lacking CAM (Table 6.2). The productivity of *C. negevensis* is apparently limited by the generally high temperature level of the Negev and by the long-lasting drought period. This could explain why *C. negevensis* is found only in shady habitats (gaps and crevices between rocks; Fig. 6.10) where full sun radiation is avoided. In this sense, it is somewhat comparable to *Dudleya arizonica* and *D. saxosa* in the southwestern United States deserts. These CAM plants occur in shady sites and grow best during the cooler season (Ting and Jennings, 1975).

Kluge and Schuber (unpublished, see also Kluge, 1978) have studied CAM in *Sedum acre* and *Sempervivum tectorum* in a natural habitat at the Botanical Gar-

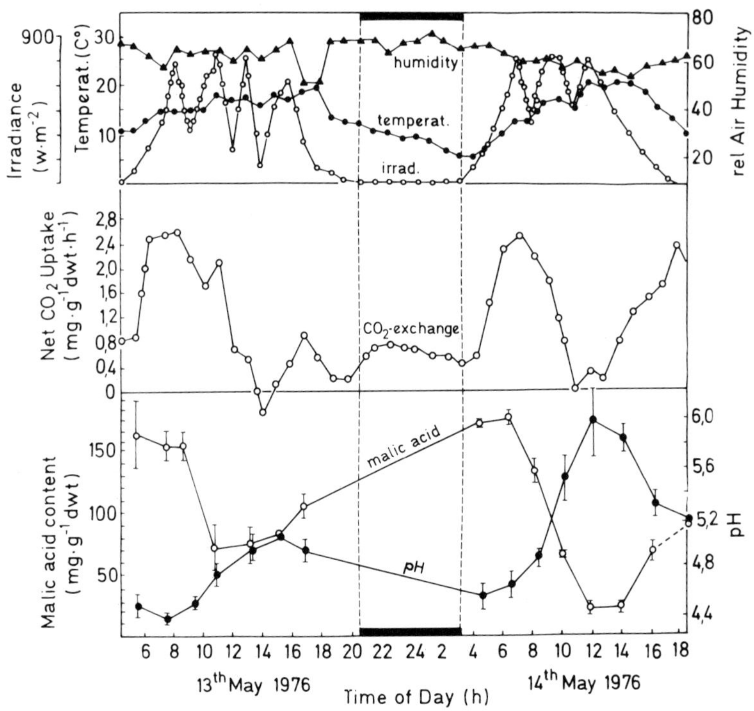

Fig. 6.11. CAM in *Sempervivum tectorum* as performed in May 1976 in its habitat at the Botanical Garden, Darmstadt (data of Kluge and Schuber; after Kluge, 1978)

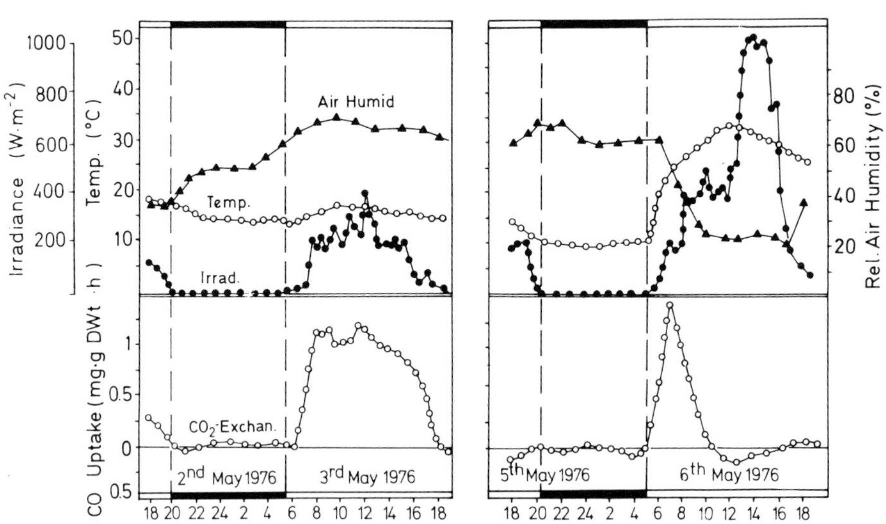

Fig. 6.12. CAM in *Sedum acre* as performed during a cool and during a hot day in May 1976 on a sand-dune habitat at the Botanical Garden, Darmstadt (data of Kluge and Schuber; after Kluge, 1978)

den in Darmstadt (Germany) using a fully climatized plant chamber and infrared gas analyzers. The δC_{13} estimates of *Sempervivum tectorum* measured by Osmond et al. (1975), suggested this species to be a CAM plant. This observation is fully supported by measurement of gas-exchange patterns and acid fluctuations (Fig. 6.11). Basically, the CO_2 exchange pattern of *Sempervivum* measured from May until August was similar to that observed in *Dudleya* as discussed above (see Fig. 6.6), with net CO_2 uptake during the night, the early hours of the day, and later afternoon. During the middle of the light period, however, sometimes net CO_2 output was observed. In experiments with the plant remaining *in situ*, the net CO_2 output has been shown to be decreased by lowering the ambient temperature. In spite of net CO_2 output during the day, carbon balance remained positive. During dark CO_2 fixation malic acid accumulation occurred, and the trough of the CO_2 uptake curve during the day was coincident with net malic acid consumption (see Kluge, 1968 a, b).

It can be expected from the data cited above that the total carbon gain in *Sempervivum tectorum* should be highest in the spring or early summer when temperatures remain relatively low both during day and night.

As shown in some preliminary control experiments, *Sempervivum tectorum* was physiologically rather inactive in the fall (end of October) with practically no net CO_2 uptake during the night, and little CO_2 uptake during the day.

CAM in *Sedum acre* is much more complicated than in *Sempervivum*. δC_{13} estimates were always in the range of −25‰ to −28‰ (Osmond et al., 1975; Kluge, 1977 b). This suggests C_3-photosynthesis rather than CAM. However, in laboratory experiments *Sedum acre* could be shifted from C_3 to CAM when water-stressed (Kluge, 1977 b). In situ, the gas exchange of *Sedum acre* showed a clear response to environmental factors (Fig. 6.12), above all by temperatures. During cool days, there was no indication of nocturnal net CO_2 fixation and malic acid accumulation. CO_2 exchange followed the C_3 type pattern and no nocturnal malic acid accumulation occurred. During May, 1976, which was unusually hot and dry, CO_2 exchange of *Sedum acre* ranged during the night near the compensation point with only a slight indication of temporary net CO_2 uptake. Malic acid increased during the night and decreased during the day, indicative of CAM. In this situation the main CO_2 uptake occurred during the early morning hours. Later in the day, CO_2 exchange changed to net CO_2 output.

These observations on *Sedum acre* suggest that CAM does not contribute considerably to net CO_2 uptake of this plant. Carbon gain should be highest in the spring with cool days and water stress situations being rare. During these periods, CO_2 is harvested via the C_3 pathway of carbon metabolism. CAM in *Sedum acre* seems to be restricted to an auxilary role as a mechanism which preserves respiratory CO_2 by nocturnal CO_2 refixation rather than contributing substantially to CO_2 uptake from the atmosphere. This may explain why, in spite of a CAM capability, *S. acre* has δC_{13} values of a C_3 plant.

6.2.2.2 Estimations of $\delta^{13}C$ values in Samples Collected in the Field

As discussed earlier in detail, in laboratory experiments, CAM in certain plant species may be shifted by environmental conditions. Water relations, temperature, and photoperiod are the major factors responsible for shifting. Based on these findings, it is reasonable to expect that a given CAM plant species growing naturally over a wide range of environmental conditions should show a relatively broad spectrum of $\delta^{13}C$ estimates in response to the environmental characteristics of the habitats.

In fact, such tendencies have been observed by Osmond (1975) who compared the $\delta^{13}C$ of the *Opuntia inermis* collected from Eastern Australia. In the more mesic southern part, *O. inermis* showed $\delta^{14}C$ values in the range of $-16.2‰$ to $-18.1‰$ indicating substantial contribution of the C_3 pathway to primary CO_2 fixation. In contrast, samples collected from more arid regions of the north and westward of the Great Dividing Range had $\delta^{13}C$ values from $-11.6‰$ to $-13.9‰$ which suggested that CO_2 was fixed entirely at night through the CAM pathway.

Similar relationships between habitat and performance of CAM have been observed by Schulze et al. (1976) in *Welwitschia mirabilis* in its natural habitat in the Namib Desert (see also Schulze and Schulze, 1976). Within the area where *Welwitschia* is naturally distributed, the $\delta^{13}C$ values were more negative than that of classical C_4 species, however, less negative than C_3 species. This indicates that *Welwitschia* in its natural habitat may fix some external CO_2 via the CAM mode of photosynthesis as certainly occurs under laboratory conditions (Dittrich and Huber, 1974). However, only in those plant individuals collected from the extremely arid coastal desert and the dry savanna were the $\delta^{13}C$ significantly less negative than that of C_3 plants. In *Welwitschia* samples obtained from the more mesic grassland, the $\delta^{13}C$ could not be distinguished from that of C_3 plants growing in the same habitat.

In spite of the above results of Osmond (1975) and Schulze et al. (1976), there is evidence from other examples of $\delta^{13}C$ analysis that under field conditions CAM is not so variable. For example, Eickmeier and Bender (1976) investigated *Agave lecheguilla* and *Opuntia engelmannii* growing along an elevational gradient at Big Bend National Park in Texas covering 1370 m. These authors showed that in CAM plants growing both over pronounced temperature and precipitation gradients, the $\delta^{13}C$ ranges in *O. engelmannii* were only 1.5‰ and 2.6‰ in *A. lecheguilla*. It was concluded that these species are obligate CAM species where the broad elevational distribution is not due to adaptional shifts in the photosynthetic option.

Szarek and Troughton (1976) studied both the mode of CO_2 fixation and $\delta^{13}C$ throughout the year in *Opuntia phaeacantha* and *Yucca baccata* along an elevational gradient from 525 m to 1900 m near Tempe (Arizona). Again, along the elevational gradient, the $\delta^{13}C$ values were similar in all plant individuals of the same species ($-12.5‰ \pm 0.86‰$ for *O. phaeacantha* and $-15.7‰ \pm 0.95‰$ for *Y. baccata*). Furthermore, despite substantial differences in temperature regimes and soil water potential, the $\delta^{13}C$ values of total tissue and soluble extracts remained remarkably constant throughout the seasons. This suggests that in *O. phaeacantha* and *Y. baccata*, CAM is always the predominating mode of photosynthesis.

Similar results were provided by Mooney et al. (1974) who collected succulents along an aridity gradient in Baja California, and in the Andes. The species of

Cactaceae and Agavaceae investigated exhibited a CAM type of $\delta^{13}C$ independent of the gradient. However, the Baja California leaf succulent *Dudleya attennata* (Crassulaceae), and Chilean Bromeliaceae in the genus *Puya* showed shifts of $\delta^{13}C$ towards less negative with increasing aridity along the gradient.

Osmond et al. (1976) investigated succulent plants from the European Alps. The values obtained suggested CAM in situ by species of *Sempervivum* (see Fig. 6.7), whereas of the Alpine *Sedum* species were about −23‰ indicating the C_3 mode dominating in CO_2 uptake.

Medina and Troughton (1974) and Medina et al. (1977) measured the $\delta^{13}C$ composition of more than 80 Bromeliaceae species. CAM as measured by CO_2 exchange and malic acid estimation was correlated with $\delta^{13}C$ estimates of −12.2‰ to −16.0‰. The ecological significance of CAM in the Bromeliaceae is suggested since the CAM species came from dry epiphytic or terrestrial habitats.

Extensive investigations of the flora on coastal and desert habitats of Israel and the Sinai were carried out by Winter et al. (1976). Only four of the 75 species studied showed diurnal malic acid rhythms typical of CAM: the halophytes *Mesembryanthemum crystallinum* (see also Winter et al., 1978), *M. forskalii* and *M. nodiflorum*, and *Caralluma negevensis* (cf. Fig. 6.5). The latter species had $\delta^{13}C$ values of −13.4‰, indicating a dominant nocturnal CO_2 fixation.

$\delta^{13}C$ estimates found in *Mesembryanthemum nodiflorum* showed differences between the specimens from Avdad (−22.0‰) in Israel and the Dead Sea (−17.2‰) which may be the result of a shift from the C_3 mode towards the CAM mode in response to different water regimes (Winter and Troughton, 1978b).

Grown under laboratory conditions, *Mesembryanthemum crystallinum* and *M. nodiflorum* (see Winter and von Willert, 1972; Treichel and Bauer, 1974) had $\delta^{13}C$ estimates of −22.7‰ and −22.0‰, which is in the range of C_3 plants, in spite of nocturnal malic acid accumulation. This suggests that the ^{13}C composition alone does not necessarily indicate CAM (see also Kluge, 1977b).

Sutton et al. (1976) irrigated *O. basilaris* plants at the natural stands for a 10-week period in order to induce a shift from CAM to C_3. By $\delta^{13}C$ analysis, there was no indication of a shift from the mean of −12‰ to −13‰ in irrigated or any control plants.

6.2.2.3 Ecological Relevance of Optional CAM

It has been outlined earlier in detail (see Chaps. 4.3.4.2 and 5.1.3.2) that under water stress a variety of plants may shift reversibly from the C_3 mode to CAM. This mechanism has been proposed to be of adaptional relevance.

Winter et al. (1978) studied the annual *Mesembryanthemum crystallinum* in its natural habitat along the mediterranean coast of Israel. As discussed in Chapter 4.3.4.2, *M. crystallinum* under laboratory conditions features CAM induction by salt (and water) stress. In its natural habitat, young plants developing from seeds during the rainy season (January and February) show C_3 photosynthesis indicated by the lack of acid oscillation and by $\delta^{13}C$ values of −25‰. During this time, the highest growth rate is observed. With the onset of the dry months, the plants became branched and produced smaller leaves. The $\delta^{13}C$ values became less negative and were about −16‰ at the termination of the life cycle. Performance of CAM during the dry season was also indicated by the diurnal oscillation of the malic acid content of the leaves. Basically, *M. nodiflorum*, also studied by Winter along the coast and in the Negev desert of Israel, showed a behavior similar to *M. crystallinum*. This reconfirms findings on *M. nodiflorum* obtained by Mooney et al. (1974).

However, in those individuals studied in the Negev desert, CAM, including its typical $\delta^{13}C$, developed earlier in the year than in coastal plants.

It is reasonable to assume that the seasonal shifting from the C_3 to the CAM mode in *Mesembryanthemum* is accelerated by the decreasing water availability. However, the above observations support also the view that ontogenetical effects influence CAM induction (von Willert et al., 1977a; Osmond, 1978).

Lange and Zuber (1977) recently discovered that *Frerea indica* Dalzell (Asclepiadaceae; = *Caralluma frerei*) changes its CO_2 metabolism by a seasonal dimorphism rather than by shifting metabolism as do the above facultative CAM plants. *Frerea indica* is a perennial stem succulent with nonsucculent leaves. The seasonal dimorphism is due to leaf abscission during the dry season. It was found that the leaves fix CO_2 via the C_3 mode; the succulent stem, however, fixes CO_2 via CAM. As a consequence of this peculiarity, together with seasonal leaf abscission, *F. indica* is able to harvest CO_2 with high photosynthetic rates during the wet season via the C_3 mode. Contrastingly, during the dry season, the plant persists in the leafless state, maintaining a good water balance by CAM.

Until now, there is no observation reported where facultative CAM plants once induced to CAM are completely reversed under field condition to the C_3 mode. This statement does not exclude, however, that the performance of induced CAM may be modulated by external factors (for example, by temperature) in the direction of non-CAM pattern.

Von Willert et al. (1978) showed CAM performance in situ in all of the 27 species of Mesembryanthemaceae growing in the Richtersveld (Southwest Africa). In this region hot desert winds may suddenly arise and increase the temperatures even during night time well above 30° C. This situation can be maintained for several days. Under these conditions the Mesembranthemaceae investigated did no longer perform CAM. Rather, they lowered continuously the malate levels until no malate was measurable. From these observations the above authors concluded that in the Mesembryanthemaceae the CAM is not capable of quick adaptation on unfavourable external conditions, e.g., high nocturnal temperatures (see also Mooney et al., 1977a).

6.2.3 Conclusions

Laboratory and field studies with CAM plants have confirmed the hypothesis that the CAM sequence is an adaptation to arid or otherwise dry environments. Thus, in the CAM-performing plant species not only are there the anatomical and morphological adaptations of the xerophytics (see Haberlandt, 1918), but also metabolic and biochemical adaptations with ecological significance.

The CAM pathway results in carbon gain with minimum water loss primarily through an efficient night CO_2-gathering pathway when the evaporation is low. Thus, stomata open without water balance being significantly affected. Hence, CAM plants are typical drought-resisters (see Levitt, 1976), which in fact resist drought by their ability to collect water during times when it is easily available, storing that water in specialized tissues during drought periods.

The results of recent field research fully support the theory that nocturnal net CO_2 fixation through the CAM pathway contributes considerably to maintaining a positive carbon balance, in particular during periods when water is in short supply. During periods of high precipitation, some CAM plants also fix substantial amounts of CO_2 during the day. Hence, the productivity of many CAM plants is highest during the wet and cool season.

Some annual CAM plants, for example *Mesembryanthemum crystallinum*, begin their life cycles in the C_3 mode, and develop CAM during the dry months of the year.

Only under extreme conditions and drought is respiratory CO_2 the sole source of carbon in CAM. Under extreme drought conditions, CAM plants may seal hermatically with virtually no water loss and CO_2 exchange. Carbon from respiration is fully recycled through CAM.

In general, the CAM photosynthetic sequence results in carbon gain rates below C_3 and C_4 plants with the result of low overall productivity. Thus, CAM is largely a survival adaptation and not, as in the case of C_4 plants, a mechanism to obtain high carbon gain rates (see Table 6.3).

6.3 Productivity

It was indicated earlier that the gas-exchange data obtained in laboratory experiments suggested that photosynthetic rates of CAM plants were considerably lower than in C_3- and C_4-species (Table 6.3). This is consistent with the general observation that CAM plants have relatively low growth rates, and supports the view that CAM is more a survival mechanism.

Because of a widespread interest in marginal dry-land agriculture, there is interest in CAM plant productivity under agricultural conditions.

Griffiths (1915) reported a series of United States Department of Agriculture-sponsored field trials for *Opuntia* production in the southwestern United States. Without cultivation, and after 2 years growth, *Opuntia sp.* produced a range from 7 to 70 tons ha^{-1} yr^{-1} field weight. Under cultivation, the same species produced from 100 to 265 tons ha^{-1} yr^{-1} field weight. In North Africa, Le-Houerou (1970) reported an average yield of 35 tons ha^{-1} yr^{-1} for *Opuntia ficus-indica;* and in North Brazil, Metral (1965) reported a yield of 102.8 tons ha^{-1} for the same species. A response to nitrogen and phosphorus was suggested.

Xolocotzi (1970) reported a yield of *Opuntia sp.* fruit of 9 tons ha^{-1} yr^{-1} with a plant density of 2000 per ha.

Growth rates of the saguaro, *Carnegia gigantea*, have been studied by Hastings and Alcon (1961). By estimating height, they found that the mean growth over the period 1951–1960 in Arizona was 6.5 cm per year, with a maximum mean growth in 1951/52 of 9.9 and a minimum in 1956/57 of 4.3 cm. The variation might be accounted for by summer precipitation.

On analysis by height class, it was found that the maximum apical growth occurs within the height class 1.2–1.8 m and falls off to 4.1 m, maintaining a low growth thereafter up to the maximum of 12 m or more.

Table 6.3. Some ecologically relevant data distinguishing CAM, C_4- and C_3-plants (data from Black, 1973)

	CAM	C_4	C_3
Transpiration ratio ($gH_2O\ g^{-1}$ dry matter)	18–100 (dark)[a] 150–600 (light)[a]	250–350	450–950
Maximum rate of net photosynthesis ($mg\ CO_2\ dm^{-2}$ leaf surface h^{-1})	1– 4 (highest reported values 11–13)	40– 80	15– 40
Maximum growth rate (g dry wt. $dm^{-2}\ d^{-1}$)	0.015–0.018	4– 5	0.5–2

[a] Data of Ting.

Although generally not available, there are many studies on production of pineapple, *Ananas comosus*. Wee (1969), for example, reported a yield of fruit of about 1.6 tons $ha^{-1}\ yr^{-1}$. From a variety of data cited by Bartholomew and Kadzimin (1977), pineapple yield varied from 8.94 to 0.81 tons ha^{-1} month^{-} with an average of about 44 tons ha^{-} yr^{-1} for the plant crop. Thus, productivity of pineapple approaches the lower limit of growth rates observed in C_3 plants (Osmond, 1978). The productivity of pineapple is exceptionally high among the CAM plants. Cultivated and wild *Opuntia*, even under irrigation, have growth rates of only about 25% that of pineapple (Osmond, 1978), and *Ferocactus* and *Agave* under desert rainfall conditions have only 8% of pineapple growth rates.

Studies of primary production by CAM plants in native environments are quite rare. At the Boyd Desert Research Center in Southern California, we estimated primary production for *Opuntia* species from 2 to 7 kg $ha^{-1}\ yr^{-1}$ and Mario Ramirez remarked at the Simposio Mundial de Zonas Arida, Mexico in 1970 that from 150 to 200 kg $ha^{-1}\ yr^{-1}$ of nopales can be gathered.

Nobel (1977b) measured the productivity of *Ferocactus acanthodes* in its natural habitat. For the one-year period of observation (8 July 1975 to 21 July 1976), the estimated net CO_2 uptake was 1.16 mmol cm^{-2} stem surface. This corresponds to 34.8 mg carbohydrate cm^{-2} or 151 g carbohydrate per plant for this period. This carbohydrate production accounted for 1.6 kg total weight (i.e., 15% of the stem mass on July 1976). These data were excellently supported by the measured increment in size during the time of observation, which was $9\% \pm 3\%$ as indicated by the measured changes in height and diameter of the cactus. Based on the above data, Nobel (1977b) calculated that a moderately large Barrel Cactus of about 90 cm tall and 35 cm in diameter would be about 54 years old.

The great capacity for growth of certain succulents under favorable conditions is exemplified by the invasion and spread of cacti throughout a large area of Queensland and New South Wales, Australia between 1890 and 1920. Several species of *Opuntia* were introduced by sea captains and other visitors as curiosities and some *Opuntia monocantha*, for example, were intended for commercial purposes. *Opuntia inermis* (see Chap. 6.2.2.1 and Fig. 6.8) became the most serious pest (Fig. 6.13), but has largely been controlled by the moth *Cactoblastus*. Presently, *Eriocereus*, the Harissia Cactus, is a serious pest of valuable rangeland in Queensland (Fig. 6.14).

Fig. 6.13a and b. *Opuntia inermis* growing in tall grass; New South Wales, Australia. **a**, 1975 photo in largely controlled area (photo by I.P.Ting). **b**, 1924 photo during peak infestation (photo from J.Mann: Cacti naturalized in Australia and their control. Dept. of Lands, Queensland; 1970, by permission)

Fig. 6.14 a and b. *Harissia* cactus growing in Queensland, Australia (Photos 1975)

Fig. 6.15. Cactus invasion in overgrazed field. Wickenberg, Arizona (Photo 1973)

In Arizona, a 17-year study indicated that *Opuntia* species have increased because of overgrazing (Glendening, 1952) (Fig. 6.15). In New Mexico, increased growth of *O. imbricata* was reported at the expense of valuable blue grama grass (Pieper, 1971), and in the Central Plains area of Colorado *O. polyacantha* is in pest proportions (Bement, 1968).

6.4 Economic Exploitation

CAM species have only been exploited on a limited basis. Without doubt, the pineapple is the most important. In Hawaii alone it represents a business, and it is extensively cultivated in Australia and Malaysia.

Opuntia spp. have variously been used in marginal agricultural areas as a supplement to cattle fodder. In the early 1900's Opuntias were used extensively, but because of the ease of cultivating other crops and the low protein value, little use is made of them now. The most commonly cultivated cacti are the flat-stemmed platyopuntias including *Opuntia ficus-indica* (throughout the world in South Africa, South America, and Mexico) and *Nopalea sp.* These are called palmas in Brazil. The edible fruits are termed tunas and the nearly spineless flat stem joints are called nopales. Both fruits and stem material are suitable for cattle food, providing a protein supplement is available. For example, in Brazil beef cattle eat and survive on freshly chopped nopales, and the nopales are mixed with cotton cakes for dairy cattle feed (Metral, 1965). For human consumption, fruit (tunas) can be eaten directly or prepared as jellies, jams, and candies. Nopales can be used raw in salads, pickled, or cooked by a variety of means.

With respect to nutrition, cacti are low in protein and high in carbohydrate. In an analysis, Martinez (1959) cites for *Opuntia ficus-indica*: water, 61.8%; fat, 2.2%; carbohydrate, 43%; protein, 5.3%; fiber, 14.3%; and ash, 28.4%.

Opuntia, in particular *Opuntia ficus-indica*, is the indirect source of cochineal (carmine lake), a red pigment which was of great practical importance until the invention of the aniline-based pigments at the beginning of this century. Even now cochineal is still used in the production of lipsticks of especially high quality. The pigment is produced by a shieldlouse, *Dactylopius coccus*, which lives on the cacti. The insects are collected, dried, and the pigment is extracted in an alcoholic solution from the powdered material. For cochineal production, *O. ficus-indica* was imported, together with its host, from Mexico to the Canary Islands, where during the 19th century the production of cochineal lac became the most important industry. Since then, *Opuntia* remained one of the dominant succulents in the vegetation of the Canaries.

Fiber obtained from *Agave* and *Yucca* is used for rope and cord. Among the important species are *Agave sisalina*, *A. fourcroydes*, *A. deweyana*, *A. lecheguilla*, and *Yucca filifera*, *Y. decipiens*, *Y. brevifolia*, and *Y. valida*. There may be some potential for cellulose production from *Yucca* (Pina, 1974).

Yucca, and *Agave*, are also potential sources of alcohol. The alcoholic beverage, *tequila*, is obtained from *Agave tequilana*, one of the most extensively cultivated agaves.

An orchid, *Vanilla*, is used economically in the West Indies and South America, and the liliaceous CAM plants are exploited for fiber and medicinal purposes. Euphorbias *(E. antisphilitica)* are used for wax production in Mexico. Aizoaceae are extensively used as ground covers.

Finally, virtually all the CAM plant families have some economic importance as ornamentals. Perhaps the most important are the Crassulaceae and the Cactaceae, but the orchids and bromeliads are extensively used.

For the future, because of their ability to produce under arid conditions, greater exploitation of CAM plants is to be expected. As indicated by Bartholomew (1975), pineapple can be cultivated without irrigation in regions with 100 cm precipitation. In Wahiawa (Hawaii) with an annual rainfall of 100 cm and a pan evaporation of 185 cm, pineapple can yield as much as 4.4 tons ha^{-1} $month^{-1}$. The potential for raw cellulose, combustable hydrocarbons, alkaloid drugs, and food and fiber production under marginal agricultural conditions is unlimited.

An example of potential use is the study by Paredes-Lopez (1974) of *Candida* growth on the juice from *Opuntia* tunas. From 100 kg fresh fruit, 3 kg of yeast cells were produced, which were 40–50% protein on a dry weight basis. As stated by these authors, the millions of tons of fruit wasted annually could be used for useful protein production.

References

Allaway, W. G.: Accumulation of malate in guard cells of Vicia faba during stomatal opening. Planta *110*, 63–70 (1973)

Allaway, W. G., Austin, B., Slatyer, R. O.: Carbon dioxide and water vapour exchange parameters of photosynthesis in a Crassulacean plant *Kalanchoe daigremontiana*. Aust. J. Plant Physiol. *1*, 397–405 (1974a)

Allaway, W. G., Osmond, C. B., Troughton, J. H.: Environmental regulation of growth, photosynthetic pathway and carbon isotope discrimination ratio in plants capable of Crassulacean acid metabolism. Proc. Int. Conf. Mechanisms of Regulation of Plant Growth. Bull. R. Soc. N.Z. *12*, 195–202 (1974b)

Ansari, A. Q., Loomis, W. E.: Leaf temperatures. Am. J. Bot. *46*, 713–717 (1959)

Aubert, B.: Etude de la resistance a la diffusion gazeuse au niveau de l'epiderme Jularie du Gananier (*Musa acuminata* Colla cv. *sinensis*) et de l'ananus (*Ananas comosus* (L.) (Merr.)) en conditions naturalles. Fruits *25*, 495–507 (1970)

Aubert, B.: Effects de la radiation globale sur la synthèse d'acides organiques et la regulation stomatique de plants succulentes exemple d'*Ananas comosus* (L.) Merr. Oecol. Plants *6*, 25–34 (1971)

Avadhani, P. N.: Studies on the fixation of carbon dioxide in leaves and roots of *Kalanchoe* and roots of *Ricinus*. Ph.D. Thesis. Newcastle: Durham Univ. 1957

Avadhani, P. N., Osmond, C. B., Tan, K. K.: Crassulacean acid metabolism and the C_4 pathway of photosynthesis in succulent plants. In: Photosynthesis and photorespiration. Hatch, M. D., Osmond, C. B., Slatyer, R. O. (eds.), pp. 288–293. New York: Wiley-Interscience 1971

Badger, M. R., Andrews, T. J., Osmond, C. B.: Detection in C_3, C_4, and CAM plant leaves of a low K_m (CO_2) form of RuDP carboxylase, having high RuDP oxygenase at physiological pH. In: Proc. 3rd int. congr. on photosynthesis. Avron, M. (ed.) pp. 1421–1429. Amsterdam: Elsevier 1975

Bahr, J. T., Jensen, R. G.: Ribulose diphosphate carboxylase from freshly ruptured spinach chloroplasts having an in vivo K_m(CO_2). Plant Physiol. *53*, 39–44 (1974)

Bandurski, R. S.: Further studies on the enzymatic synthesis of oxaloacetate from phosphoenolpyruvate and carbon dioxide. J. Biol. Chem. *217*, 137–150 (1955)

Bandurski, R. S., Greiner, C. M.: The enzymatic synthesis of oxaloacetate from phosphoenolpyruvate and carbon dioxide. J. Biol. Chem. *204*, 781–786 (1953)

Bartholomew, B.: Drought response in the gas exchange of *Dudleya farinosa (Crassulaceae)* grown under natural conditions. Photosynthetica 7, 114–120 (1973)

Bartholomew, D. P.: Use of diffusion porometer on xerophytic plants. In: Measurement of stomatal aperture and diffusion resistance. Kanemasu, E. T. (ed.). Wash. Agric. Exp. Stu. Bull. 1975

Bartholomew, D. P., Kadzimin, S. B.: Pineapple. In: Ecophysiology of tropical crops. Alvin, P. T. (ed.). Brazil: CEPLAC 1977

Baskin, J. M., Baskin, C. C.: Leaf temperature of *Agave virginica* (L.) during a cloudy and a clear day in summer. J. Tenn. Acad. Sci. *46*, 141–142 (1971)

Baskin, J. M., Baskin, C. C.: Pod temperatures of *Opuntia compressa* during daytime in summer. Bull. Torrey Bot. Club *100*, 56–59 (1973)

Beevers, H., Gibbs, M.: The direct oxydation pathway in plant respiration. Plant Physiol. *29*, 322–324 (1954)

Beevers, H., Stiller, M. L., Butt, V. S.: Metabolism of the organic acids. In: Plant physiol. Stewart, F. C. (ed.), Vol. 4, pp. 119–262. New York: Academic Press 1966

Bement, R. E.: Plains prickly pear: relation to grazing intensity and blue grama yield in central Great Plains. J. Range Manage *21*, 83–86 (1968)

Bender, M. M.: Mass spectrometric studies of carbon 13 variation in corn and other grasses. Radiocarbon *10*, 468–472 (1968)

Bender, M. M.: Variations in $^{13}C/^{12}C$ ratios of plants in relation to the pathway of photosynthetic carbon dioxide fixation. Phytochem. *10*, 1239–1244 (1971)

Bennet-Clark, T. A.: The role of organic acids in plant metabolism. Part I. New Phytol. *32*, 37–71 (1933a)

Bennet-Clark, T. A.: The role of organic acids in plant metabolism. Part II. New Phytol. *32*, 128–161 (1933b)

Bennet-Clark, T. A.: Organic acids of plants. Ann. Rev. Biochem. *6*, 579–594 (1937)

Bennet-Clark, T. A.: Organic acids of plants. Ann. Rev. Biochem. *18*, 639–654 (1949)

Benzing, D. H., Renfrow, A.: Significance of the pattern of CO_2 exchange to the ecology and phylogeny of the *Tillandsioideae (Bromeliaceae)*. Bull. Torrey Bot. Club *98*, 322–327 (1971)

Bernatek, E., Nordal, A., Ogner, G.: Phorbic acid, a new acid from *Euphorba*. Acta Chem. Scand. *17*, 2375–2384 (1963)

Bernstein, L.: Osmotic adjustements of plants to saline media. II. Dynamic phase. Am. J. Bot. *50*, 360–370 (1963)

Bierhuizen, J. A., Slatyer, R. O.: Photosynthesis of cotton leaves under a range of environmental conditions in relation to internal and external diffusive resistance. Aust. J. Biol. Sci. *17*, 348–359 (1964)

Björkman, O., Osmond, C. B.: Effect of oxygen on carbon dioxide fixation in *Kalanchoe daigremontiana*. Carnegie Inst. Yearb. *73*, 852–858 (1974)

Björkman, O., Pearcy, W., Harrison, A. T., Mooney, H. A.: Photosynthetic adaptation to high temperatures: a field study in Death Valley, California. Science *175*, 786–789 (1972)

Black, C. C.: Photosynthetic carbon fixation in relation to net CO_2 uptake. Ann. Rev. Plant Physiol. *24*, 253–286 (1973)

Black, C. C., Campbell, W. H., Chen, T. M., Dittrich, P.: The monocotyledones: their evolution and comparative biology. III. Pathways of carbon metabolism related to net carbon dioxide assimilation by monocotyledons. Q. Rev. Biol. *48*, 299–313 (1973)

Black, C. C., Williams, S.: Plants exhibiting characteristics common to Crassulacean Acid Metabolism. In: CO_2 metabolism and plant productivity. Burris, R. H., Black, C. C. (eds.), pp. 407–424. Baltimore: Univ. Park Press 1976

Bode, O.: Über Zusammenhänge zwischen CO_2-Assimilation und Photoperiodismus bei *Kalanchoe blossfeldiana*. Planta *33*, 278–289 (1942)

Böcher, M., Kluge, M.: The C_4-pathway of C-fixation in *Spinacea oleracea*. I. ^{14}C-labelling pattern of suspended leaf slices as influenced by the external medium. Z. Pflanzenphysiol. *83*, 347–361 (1977)

Böcher, M., Kluge, M.: The C_4-pathway of C-fixation in *Spinacea oleracea*. II. Pulse chase experiments with suspended leaf slices. Z. Pflanzenphysiol. *86*, 405–421 (1978)

Bonner, W., Bonner, J.: The role of carbon dioxide in acid formation by succulent plants. Am. J. Bot. *35*, 113–117 (1948)

Bowes, G., Ogren, W. L.: O_2 inhibition and other properties of soybean ribulose 1,5-diphosphate carboxylase. J. Biol. Chem. *247*, 2171–2176 (1972)

Bowling, D. J. F.: Malate-switch hypothesis to explain the action of stomata. Nature (London) *262*, 393–394 (1976)

Bradbeer, J. W.: Some carbohydrate and acid conversions in *Kalanchoe crenata*. Doctoral Thesis, Durham Univ. Durham, England (Newcastle Division) 1954

Bradbeer, J. W., Ranson, S. L.: Physiological studies on acid metabolism in green plants. VIII. The utilization of labelled pyruvate, fumarate, and glucose in *Kalanchoe* leaves in the dark. Proc. R. Soc. *157 B*, 258–278 (1963)

Bradbeer, J. W., Ranson, S. L., Stiller, M.: Malate synthesis in Crassulacean leaves. I. The distribution of ^{14}C in malate of leaves exposed in $^{14}CO_2$ in the dark. Plant Physiol. *33*, 66–70 (1958)

Brandon, P. C.: Temperature features of enzymes affecting crassulacean acid metabolism. Plant Physiol. *42*, 977–984 (1967)

Britton, N. L., Rose, J. N.: The *Cactaceae*: Descriptions and illustrations of the plants of the cactus family in four volumes bound as two. I and II–III and IV. New York: Dover Publications, Inc. 1963

Brown, J. M. A.: Studies of inhibitors in relation to crassulacean acid metabolism. Doctoral thesis, Durham Univ. Durham England (Newcastle Division) 1956

Brownwell, P. F., Crossland, C. J.: Sodium requirement in relation to the photosynthetic options of a CAM plant. Plant Physiol. *54*, 415–417 (1974)

Bruinsma, J.: Studies on the crassulacean acid metabolism. Acta Bot. Neerl. *7*, 531–588 (1958)

Brunnhöfer, H., Schaub, H., Egle, K.: Der Verlauf des CO_2 und O_2 Gaswechsels bei *Bryophyllum daigremontianum* in Abhängigkeit von der Temperatur. Z. Pflanzenphysiol. *59*, 285–292 (1968a)

Brunnhöfer, H., Schaub, H., Egle, K.: Die Beziehungen zwischen den Veränderungen der Malat- und Stärkekonzentration und dem CO_2- und O_2-Gaswechsel bei *Bryophyllum daigremontianum*. Z. Pflanzenphysiol. *60*, 12–18 (1968b)

Bull, T. A.: Photosynthetic efficiencies and photorespiration in Calvin cycle and C_4 dicarboxylic acid plants. Copper Sci. *9*, 726–729 (1969)

Bünning, E.: The physiological clock. The English Univ. Press London. Heidelberg, New York: Springer 1973

Burris, R. H.: Organic acids in plant metabolism. Ann. Rev. Plant Physiol. *4*, 91–114 (1953)

Buser, Ch., Matile, Ph.: Malic acid in vacuoles isolated from *Bryophyllum* leaf cells. Z. Pflanzenphysiol. *82*, 462–466 (1977)

Calvin, M., Bassham, J. A., Benson, A. A., Lynch, V. H., Ovellet, C., Schou, L., Stepka, W., Tolbert, N. E.: Carbon dioxide assimilation in plants. Symp. Soc. Exp. Biol. *5*, 284–305 (1951)

Canovas, J. L., Kornberg, H. L.: Properties and regulation of phosphoenolpyruvate carboxylase activity in *Escherichia coli*. Proc. R. Soc. Ser. B Biol. Sci. *165*, 180–205 (1966)

Champigny, M. L.: Lumière et genèse des acides amines. Ph.D. Thesis. University Paris 1960

Champigny, M. L., Jolchine, G., Moyse, A.: Sur la distribution de $^{14}CO_2$ fixe photosynthese et par *β*-carboxylation dans les feuilles de *Bryophyllum*. Conf. Int. Radioisotopes Rech. Sc. UNESCO, pp. 458–470. Paris: Pergamon Press 1958

Chaturvedi, S. N., Zabka, G.: Studies on dark fixation of carbon dioxide in *Kalanchoë*. I. Effect of water stress and growth retardants. Ann. Bot. *41*, 493–500 (1977a)

Chaturvedi, S. N., Zabka, G.: Studies on dark fixation of carbon dioxide in *Kalanchoe*. II. Effect of interaction of photoperiodic induction with water stress and growth retardants. Ann. Bot. *41*, 501–505 (1977b)

Chesnokov, V. A., Miroslavova, S. A.: The exchange of malic acid in the light in leaves of succulents. Leningrad Univ. Herald [Russian] *21*, 141–152 (1969)

Chow, P. N., Burnside, O. C., Lavy, T. L.: Physiological studies with prickly pear. Weeds *14*, 58–62 (1966)

Chu, D. K., Bassham, J. A.: Activation of ribulose-1,5-diphosphate carboxylase by nicotinamide adenine dinucleotide phosphate and other chloroplast metabolites. Plant Physiol. *54*, 556–559 (1974)

Cockburn, W.: Problems relating to acid synthesis in higher green plants. Ph.D. Thesis. England: Univ. of Newcastle on Tyne 1965

Cockburn, W.: Crassulacean acid metabolism in *Lithops insularis*, a non-halophytic member of the Mesembryanthemaceae. Planta *118*, 89–90 (1974)

Cockburn, W., McAulay, A.: The pathway of malate synthesis in crassulacean acid metabolism. In: Environmental and biological control of photosynthesis. Marcelle, R., Junk, Dr. W. b.v. Publ. The Hague 1975a

Cockburn, W., McAulay, A.: The pathway of carbon dioxide fixation in crassulacean plants. Plant Physiol. 55, 87–89 (1975b)

Cockburn, W., McAulay, A.: Changes in metabolite levels in *Kalanchoe daigremontiana* and regulation of malic acid accumulation in crassulacean acid metabolism. Plant Physiol. *59*(3), 455–458 (1977)

Conde, L. F., Kramer, P. J.: The effect of vapor pressure deficit and diffusion resistance in *Opuntia compressa*. Can. J. Bot. *53*, 2923–2926 (1975)

Coombs, J., Baldry, C. W., Bucke, C.: The C_4 pathway in *Pennisetum purpureum*. I. The allosteric nature of PEP carboxylase. Planta *110*, 95–107 (1973)

Coombs, J., Maw, S.L., Baldry, C.W.: Metabolic regulation in C_4 photosynthesis; PEP carboxylase and energy charge. Planta *117*, 279–292 (1975)
Cooper, T.G., Filmer, O., Wischnick, M., Lane, M.D.: The active species of "CO_2" utilized by ribulose diphosphate carboxylase. J. Biol. Chem. *244*, 1081 (1969)
Cooper, T.G., Tchen, T.T., Wood, H.G., Benedict, C.R.: The carboxylation of phosphoenolpyruvate and pyruvate. I. The active species of "CO_2" utilized by phosphoenolpyruvate carboxykinase, carboxytransphosphorylase, and pyruvate carboxylase. J. Biol. Chem. *243*, 3857–3863 (1968)
Cooper, T.G., Wood, H.G.: The carboxylation of phosphoenolpyruvate carboxylase. J. Biol. Chem. *246*, 5488–5490 (1971)
Coutinho, L.M.: Algumus informacoes sobre a ocorenica do "efeito De Saussure" em epifitas e erbaceas terrestres de mata pluvial. Botanica *20*, 83–98 (1963)
Coutinho, L.M.: Untersuchungen über die Lage der Lichtkompensationspunkte einiger Pflanzen zu verschiedenen Zeiten mit besonderer Berücksichtigung des „De Saussure Effekts" bei Sukkulenten. Beitr. Phytologie (Walter-Festschrift), S. 101–108. Stuttgart 1964
Coutinho, L.M.: Alguma informacoes sobre a capacidadae ritmica diaria da fixacao e acumulacae de CO_2 no escuro em epifitas e erbacias terrestre da mata pluvial. Botanica *21*, 397–408 (1965)
Coutinho, L.M.: Novas observacoes sobre a ocurrencia do "efeito de Saussure" e suas relacoes com a susulencia, a temperature folhear e os movimentos estimaticos. Botanica *24*, 44–102 (1969)
Crews, C.E., Vines, H.M., Black, C.C.: Post illumination burst of carbon dioxide in crassulacean acid metabolism plants. Plant Physiol. *55*, 652–657 (1975)
Crews, C.E., Williams, S.L., Vines, H.M., Black, C.C.: Changes in the metabolism and physiology of crassulacean acid metabolism plants grown in controlled environments. In: CO_2 metabolism and plant productivity. Burris, R.H., Black, C.C. (eds.), pp. 235–250. Baltimore, London, Tokyo: University Park Press 1976
Daley, L.S., Ray, T., Vines, H.M., Black, C.C.: Pineapple PEP carboxykinase: partial purification and properties. Plant Physiol. *57*(5) Suppl. 32 (1976)
Danner, J., Ting, I.P.: CO_2 metabolism in corn roots. II. Intracellular distribution of enzymes. Plant Physiol. *42*, 719–724 (1967)
Davies, D.D.: Organic acid metabolism in plants. Biol. Rev. *34*, 407–444 (1959)
Dayanandan, P., Kaufman, P.B.: Stomatal movements associated with potassium fluxes. Am. J. Bot. *62*, 221–231 (1975)
Delf, E.M.: Transpiration in succulent plants. Ann. Bot. *26*, 409–440 (1912)
Deleens, E., Garnier-Dardart, J.: Carbon isotope composition of biochemical fractions isolated from leaves of *Bryophyllum daigremontianum* Berger, a plant with Crassulacean Acid Metabolism: Some physiological aspects related to CO_2 dark fixation. Planta *139*, 214–248 (1977)
Deleens, E., Lerman, J.C., Nato, A., Moyse, A.: Carbon isotope discrimination by the carboxylating reactions in C_3, C_4, and CAM plants. Proc. 3rd Int. Congr. Photosynth., *2*. Avron, M. (ed.), pp. 1267–1276. Amsterdam: Elsevier 1974
Denius, H.R., Homann, P.: The relation between photosynthesis, respiration and Crassulacean Acid Metabolism in leaf slices of *Aloe arborescens* Mill. Plant Physiol. *49*, 873–880 (1972)
De Saussure, T.: Recherches chimiques sur la vegetation. Paris: Nyon 1804. Quoted in Wolf, 1960
Despain, D.G., Bliss, L.C., Boyer, J.S.: Carbon dioxide exchange in saguaro seedlings. Ecology *51*, 912–914 (1970)
De Vries, H.: Über die periodische Säurebildung der Fettpflanzen. Bot. Zeitung *42*, 339–344 (1884)
Dilley, D.R.: Purification and properties of apple fruit malic enzyme. Plant Physiol. *41*, 214–220 (1966)
Dinger, B.E., Patten, D.T.: Carbon dioxide exchange in selected species of *Echinocereus* (Cactaceae). Photosynthetica *6*, 345–353 (1972)
Dinger, B.E., Patten, D.T.: Carbon dioxide exchange and transpiration in species of *Echinocereus* (Cactaceae), as related to their distribution within the Pinaleno Mountains, Arizona. Oecologia *14*, 389–411 (1974)

Dittrich, P.: Nicotinamide adenine dinucleotide specific "malic" enzyme in *Kalanchoe daigremontiana* and other plants exhibiting Crassulacean Acid Metabolism. Plant Physiol. *57*, 310–314 (1975)

Dittrich, P.: Equilibrium of label in malate during dark fixation of CO_2 in *Kalanchoe fedtschenkoi*. Plant Physiol. *58*, 288–291 (1976)

Dittrich, P., Campbell, W.H., Black, C.C.Jr.: Phosphoenolpyruvate carboxykinase in plants exhibiting Crassulacean Acid Metabolism. Plant Physiol. *52*, 357–361 (1973)

Dittrich, P., Huber, W.: Carbon dioxide metabolism in members of the *Chlamydospermae*. Int. Congr. Photosynth. Avron, M. (ed.), pp. 1573–1578. Amsterdam, Netherlands: Elsevier Sci. Publ. 1974

Downton, W.J.S.: Adaptive and evolutionary aspects of C_4 photosynthesis. In: Photosynthesis and photorespiration. Hatch, M.D., Osmond, C.B., Slatyer, R.O. (eds.), pp. 3–17. New York, London, Sydney, Toronto: Wiley-Interscience 1971

Eickmeier, W.G., Bender, M.M.: Carbon isotope ratios of Crassulacean Acid Metabolism species in relation to climate and phytosociology. Oecologia *25*, 341–347 (1976)

Enama, M.: Molecular weight variations of phosphoenolpyruvate carboxylase from C_4 plants. Carnegie Inst. Yearb. *75*, 409–413 (1976)

Erler, W.L.: Daytime stomatal closure in *Agave americana* as related to enhanced wateruse efficiency. Physiological systems in semi acid environments. Hoff, C.C., Rudesch, M.L. (eds.), pp. 239–247. New Mexico: Univ. New Mexico, Press Albuquerque 1969

Evans, H.: The physiology of succulent plants. Biol. Rev. *7*, 181–211 (1932)

Fasella, P., Bossa, F., Turano, C., Fanelli, A.R.: Multiple forms of aspartate oxoglutarate transaminase and malate dehydrogenase in resting cotton seeds. Enzymologia *30*, 180–205 (1966)

Filmer, D.L., Cooper, T.G.: Effect of varying temperature and pH upon the predicted rate of "CO_2" utilization by carboxylases. J. Theor. Biol. *29*, 131–145 (1970)

Fischer, R.A.: Stomatal opening: role of potassium uptake by guard cells. Science *160*, 784–785 (1968)

Fujino, M.: Role of adenosinetriphosphate and adenosinetriphosphate in stomatal movement. Sci. Bull. Fac. Educ. Nagasaki Univ. *18*, 1–47 (1967)

Fujino, M.: Effect of monoiodoacetic acid on stomatal movement. Sci. Bull. Fac. Educ. Nagasaki Univ. *20*, 57–66 (1969)

Gaffron, H., Fager, F.W.: The kinetics and chemistry of photosynthesis. Ann. Rev. Plant Physiol. *2*, 87–114 (1951)

Ganzmann, R.J., v. Willert, D.J.: Nachweis eines diurnalen Säurerhythmus beim Halophyten *Aster tripolium*. Naturwissenschaften *59*, 422–423 (1973)

Garnier-Dardat, J.: Activités enzymatiques des chloroplastes isolès de geuilles de *Bryophyllum daigremontianum* Berger: oxydation des hexoses, formation et degradation d'acide malique. Physiol. Veg. *3*, 215–227 (1965)

Garnier-Dardart, J., Queiroz, O.: Malic enzymes in the leaves of *Bryophyllum daigremontianum*. Phytochemistry *13*, 1695–1702 (1974)

Gates, D.M.: Energy exchange in the biosphere. New York: Harper and Row. Monographs 1962

Gates, D.M.: Energy exchange and transpiration. In: Water and plant life. Ecological Studies Vol. 19. Lange, O.C., Kappen, L., Schulze, E.D. (eds.), pp. 137–147. Berlin, Heidelberg, New York: Springer 1976

Gates, D.M., Benedict, Ch.M.: Convection phenomena from plants in still air. Am. J. Bot. *50*, 563–573 (1963)

Gaur, Y.D.: Preliminary studies on titratable acidity in xerophytic plants: *Salvadora persica* Linn. and *Prosopis juliflora* D.C. Experientia *24*, 239–240 (1968)

Gibbs, J.G., Patten, D.T.: Plant temperatures and heat flux in a Sonoran desert ecosystem. Oecologia 5, 165–184 (1970)

Gibbs, M.: Effect of light intensity on the distribution of ^{14}C in sunflower leaf metabolites during photosynthesis. Arch. Biochem. Biophys. *45*, 156–160 (1953)

Gifford, E.M., Steward, K.D.: Inclusions of the proplastids and vacuoles in the shoot apices of *Bryophyllum* and *Kalanchoe*. Am. J. Bot. *55*, 269–279 (1968)

Gifford, R. M.: A comparison of potential photosynthesis productivity and yield of plant species with differing photosynthetic metabolism. Aust. J. Plant Physiol. *1*, 107–117 (1974)
Glendening, G. E.: Some quantitative data of the increase of mesquite and cactus on a desert grassland range on Southern Arizone. Ecology *32*, 319–328 (1952)
Gregory, F. G., Spear, J., Thimann, K. V.: The interrelativ between CO_2 Metabolism and photoperiodism in *Kalanchoe*. Plant Physiol. *29*, 220–229 (1954)
Griffiths, D.: Yields on native prickly pear in Southern Texas. U.S.D.A. Bull. *208*, 1–11 (1915)
Haberlandt, G.: Physiologische Pflanzenanatomie. Verlag W. Engelmann 1918
Hadley, E. B.: Net carbon dioxide exchange in succulents. Two *Kalanchoe* species and their interspecific hybrid. Proc. N.D. Acad. Sci. *19*, 170–178 (1965)
Haidri, Z. A.: Incorporation of malic acid-2-^{14}C and pyruvic acid-2-^{14}C into starch. Plant Physiol. *30*, 4 (1955a)
Haidri, Z. A.: Synthesis of starch from organic acids in plants. Ph.D. Thesis. Univ. of Wsconsin (1955b)
Hall, A. E., Kaufmann, M. R.: Stomatal response to environment with *Sesamum indicum* L. Plant Physiol. *55*, 455–459 (1975)
Hall, A. E., Schulze, E. D., Lange, O. L.: Current perspectives of steady-state stomatal responses to environment. In: Water and plant life. Problems and modern approaches. Ecol. Studies Vol. 19. Lange, O. L., Kappen, K., Schulze, E. D. (eds.), pp. 169–188. Berlin, Heidelberg, New York: Springer 1976
Hanscom, Z., Ting, I. P.: Physiological responses to irrigation in *Opuntia basilaris* Engelm. & Bigel. Bot. Gaz. *138*, 159–167 (1977)
Hanscom, Z., Ting, I. P.: Responses of succulents to plant water stress. Plant. Physiol., *61*, 327–330 (1978)
Harary, I., Korey, S. R., Ochoa, S.: Biosynthesis of dicarboxylic acids by carbon dioxide fixation. VII. Equilibrium of "malic" enzyme reaction. J. Biol. Chem. *203*, 595–604 (1953)
Harris, P. J. C., Wilkins, M. B.: Light induced changes in the period of the circadian rhythm of carbon dioxide output in *Bryophyllum* leaves. Planta *129*, 253–258 (1976)
Harris, P. J. C., Wilkins, M. B.: Evidence of phytochrom involvement in the entrainment of the circadian rhythm of carbon dioxide metabolism in *Bryophyllum*. Planta *138*, 271–278 (1978)
Hartsock, T. L., Nobel, P. S.: Watering converts a CAM plant to daytime CO_2 uptake. Nature (London) *262*, 574–576 (1976)
Hastings, J. R., Alcon, S. M.: Physical determinations of growth and age in the grand cactus. J. Ariz. Acad. Sci. *2*, 32–39 (1961)
Hatch, M. D.: Synthesis of L-malate-4-^{14}C and determination of label in the C_4 carboxyl of L-malate. Anal. Biochem. *47*, 174–183 (1972)
Hatch, M. D.: The C_4 pathway of photosynthesis: Mechanism and function. In: CO_2 metabolism and plant productivity. Burris, R. H., Black, C. C. (eds.), pp. 59–82. Baltimore, London, Tokyo: University Park Press 1976
Hatch, M. D., Osmond, C. B.: Compartmentation and transport in C_4 photosynthesis. In: Transport in plants III. Enzyclopedia of plant physiology, New series, Vol. 3. Stocking, C. R., Heber, U. (eds.), pp. 144–184. Berlin, Heidelberg, New York: Springer 1976
Hatch, M. D., Slack, C. R.: Studies on the mechanism of activation and inactivation of pyruvate, phosphate dikinase. A possible regulatory role for the enzyme in the C_4 dicarboxylic acid pathway of photosynthesis. Biochem. J. *112*, 549–558 (1969)
Hatch, M. D., Slack, C. R.: Photosynthetic CO_2 fixation pathways. Ann. Rev. Plant Physiol. *21*, 141–162 (1970)
Heber, U. W., Santarius, K. A.: Compartmentation and reduction of pyridine nucleotides in relation to photosynthesis. Biochem. Biophys. Acta *109*, 390–408 (1965)
Henckel, P. A.: Physiology of plants under drought. Ann. Rev. Plant Physiol. *15*, 363–386 (1964)
Heyne, B.: On the deoxidation of the leaves of Cotyledon calycina. Trans. Linn. Soc. Lond. *11*, p. II, 213–215 (1815)
Hew, S. C., Wong, Y. S.: Photosynthesis and respiration in ferns in relation to their habitats. Am. Fern J. *64*, 40–48 (1974)
Holdsworth, M.: Carbon dioxide uptake by succulents. Can. J. Bot. *49*, 1520–1522 (1971)

Huber, B.: Der Wärmehaushalt der Pflanzen. Naturwiss. und Landwirtschaft, Heft *17*. Freising-München: Verl. Dr. F. P. Datterer u. Cie 1935

Huber, W., Sankhla, N.: C_4 pathway and regulation of the balance between C_4 and C_3 metabolism. In: Water and plant life. Problems and modern approaches. Ecol. Studies, Vol. *19*. Lange, O. L., Kappen, L., Schulze, E. D. (eds.), pp. 335–363. Berlin, Heidelberg, New York: Springer 1976

Itai, C., Benzioni, A.: Water stress and hormonal response. In: Water and plant life. Ecological Studies, Vol. 19. Lange, O. L., Kappen, L., Schulze, E. D. (eds.), pp. 225–240. Berlin, Heidelberg, New York: 1976

Jacobsen, H.: A handbook of succulent plants. London: Bradford Press 1960

Jacoby, B., Laties, G. G.: Bicarbonate fixation and malate synthesis in relation to salt induced stoichiometric synthesis of organic acids. Plant Physiol. *47*, 525–531 (1971)

James, W. O.: Succulent plants. Endeavour. *17*, 90–95 (1958)

Johnson, E. J., Bruff, B. S.: Chloroplast integrity and ATP-dependant CO_2 fixation in *Spinacia oleracea*. Plant Physiol. *42*, 1312–1328 (1967)

Johnson, H. S., Hatch, M. D.: Properties and regulation of leaf nicotinamide-adenine dinucleotide phosphate-malate dehydrogenase and "malic" enzyme in plants with the C_4-dicarboxylic acid pathway of photosynthesis. Biochem. J. *119*, 173–280 (1970)

Jones, M. B.: Some observations on a circadian rhythm in carbon dioxide compensation in *Bryophyllum fedtschenkoi*. Ann. Bot. *37*, 1027–1034 (1973)

Jones, M. B.: The effect of leaf age on leaf resistance and CO_2 exchange of the CAM plant *Bryophyllum fedtschenkoi*. Planta *123*, 91–96 (1975)

Jones, M. B., Mansfield, T. A.: A circadian rhythm in the level of the carbon dioxide compensation point in *Bryophyllum* and *Coffea*. J. Exp. Bot. *21*, 159–163 (1970)

Jones, M. B., Mansfield, T. A.: A circadian rhythm in the level of carbon dioxide compensation in *Bryophyllum fedtschenkoi* with zero values during the transient. Planta *103*, 134–146 (1972)

Joshi, G., Dolan, T., Gee, R., Saltman, P.: Sodium chloride effect in dark fixation of CO_2 by marine and terrestrial plants. Plant Physiol. *37*, 446 (1962)

Joshi, G. V., Bartakke, S. P.: Light and dark $^{14}CO_2$ fixation in *Aloe vera* var. *indica*. Maharashtra Vidnyan Mandir Patrika *9*, 22–28 (1974)

Joshi, M. C., Boyer, J. S., Kramer, P. J.: Growth, carbon dioxide exchange, transpiration and transpiration ratio of pineapple. Bot. Gaz. *126*, 174–179 (1965)

Kanai, R. H., Black, C. C.: Biochemical basis for net CO_2 assimilation in C_4 plants. In: Net carbon dioxide assimilation in higher plants. Proc. Soc. Sect. Am. Soc. Plant Physiol. Black, C. C. (ed.), pp. 75–93. Releigh: Cotton-Inc. 1972

Kapil, R. N., Pugh, T. D., Newcomb, E. H.: Microbodies and anomalous "microcylinder" in the ultrastructure of plants with Crassulacean Acid Metabolism. Planta *124*, 231–244 (1975)

Kaplan, A., Gale, J.: Separation of respiration from dark fixation of carbon dioxide in CAM plants. Isr. J. Bot. *25*, 59 (1975)

Kaplan, A., Gale, J., Poljakoff-Mayber, A.: Simultaneous measurement of oxygen, carbon dioxide, and water vapour exchange of intact plants. J. Exp. Bot. *97*, 214–219 (1976a)

Kaplan, A., Gale, J., Poljakoff-Mayber, A.: Resolution of net dark fixation of carbon dioxide into its respiration and gross fixation components in *Bryophyllum daigremontianum*. J. Exp. Bot. *97*, 220–230 (1976b)

Kaplan, A., Gale, J., Poljakoff-Mayber, A.: Effect of oxygen and carbon dioxide concentrations on gross dark CO_2 fixation and dark respiration in *Bryophyllum daigremontianum*. Aust. J. Plant Physiol. *4*, 745–752 (1977)

Kaul, R. B.: The role of the multiple epidermis in foliar succulence of *Peperomia (Piperaceae)*. Bot. Gaz. *138*, 213–218 (1977)

Kausch, W.: Beziehungen zwischen Wurzelwachstum, Transpiration und CO_2-Gaswechsel bei einigen Kakteen. Planta *66*, 229–238 (1965)

Kelly, G. J., Latzko, E., Gibbs, M.: Regulatory aspects of photosynthetic carbon metabolism. Ann. Rev. Plant Physiol. *27*, 181–206 (1976)

Kelly, G. J., Turner, J. F.: The regulation of pea-seed phosphofructokinase by phosphoenolpyruvate. Biochem. J. *115*, 481–487 (1969)

Khan, A. A.: Oxidative metabolism of cactus plant. Ph.D. Thesis. Lucknow: Lucknow Univ. 1969
Kinraide, T. B., Behan, M. J.: Restoration of organic acid accumulation in sectioned leaves of *Bryophyllum tubiflorum* Hare. Plant Physiol. *56*, 830–835 (1975)
Kluge, M.: Untersuchungen über den Gaswechsel von *Bryophyllum* während der Lichtperiode. I. Zum Problem der CO_2-Abgabe. Planta *80*, 255–263 (1968 a)
Kluge, M.: Untersuchungen über den Gaswechsel von *Bryophyllum* während der Lichtperiode. II. Beziehungen zwischen dem Malatgehalt des Gewebes und der CO_2-Aufnahme. Planta *80*, 359–377 (1968 b)
Kluge, M.: Zur Analyse des CO_2-Austauschs von *Bryophyllum*. I. Messung der Änderung des Mengenverhältnisses einiger Phosphatverbindungen im Blattgewebe während bestimmter Phasen der Licht-Dunkel-Periode. Planta *85*, 160–170 (1969 a)
Kluge, M.: Zur Analyse des CO_2-Austausches von *Bryophyllum*. II. Hemmung des nächtlichen Stärkeabbaus in CO_2 verarmter Atmosphäre. Planta *86*, 142–150 (1969 b)
Kluge, M.: Veränderliche Markierungsmuster bei $^{14}CO_2$-Fütterung von *Bryophyllum tubiflorum* zu verschiedenen Zeitpunkten der Hell-Dunkel-Periode. I. Die Fütterung unter Belichtung. Planta *88*, 113–129 (1969 c)
Kluge, M.: Studies on CO_2 fixation by succulent plants in the light. In: Photosynthesis and photorespiration. Hatch, M. D., Osmond, C. B., Slatyer, R. O. (eds.), pp. 283–287. New York: Wiley-Interscience 1971 a
Kluge, M.: Veränderliche Markierungsmuster bei $^{14}CO_2$-Fütterung von *Bryophyllum tubiflorum* zu verschiedenen Zeitpunkten der Hell-Dunkel-Periode. II. Beziehungen zwischen dem Malatgehalt des Gewebes und dem Markierungsmuster nach $^{14}CO_2$-Lichtfixierung. Planta *98*, 20–30 (1971 b)
Kluge, M.: Der CO_2-Austausch der Sukkulenten: Biochemische Grundlagen einer ökologischen Anpassung. Ber. Dtsch. Bot. Ges. *84*, 417–424 (1971 c)
Kluge, M.: Crassulacean acid metabolism (CAM): CO_2 and water economy. In: Water and plant life. Problems and modern approaches. Ecological Studies, Vol. 19. Lange, O. L., Kappen, L., Schulze, E. D. (eds.), pp. 313–323. 1976 a
Kluge, M.: Models of CAM regulation. In: CO_2 metabolism and plant productivity. Burris, R. H., Black, C. C. (eds.), pp. 205–216. Baltimore, London, Tokyo: University Park Press 1976 b
Kluge, M.: Regulation of carbon dioxide fixation in plants. In: Integration of Activity in the Higher Plant. Symposium of Soc. Exp. Biol., Vol. XXXI. Jennings, D. H. (ed.), pp. 155–175. Cambridge, New York, Melbourne: Cambridge University Press 1977 a
Kluge, W.: Is *Sedum acre* a CAM plant? Oecologia *29*, 77–83 (1977 b)
Kluge, M.: Ecological aspects of crassulacean acid metabolism (CAM). In: Photosynthesis 77: Proc. of 4th Internat. Congress of Photosynthesis. Hall, D. O., Coombs, J., Goodwin, T. W. (eds.), pp. 335–344. London: Biochem. Soc. 1978
Kluge, M., Bley, L., Schmid, R.: Malate synthesis in Crassulacean Acid Metabolism (CAM) via a double CO_2 dark fixation? In: Environmental and biological control of photosynthesis. Marcelle, R. (ed.), pp. 281–288. The Hague: Dr. W. Jung b.v. Publishers, 1975
Kluge, M., Fischer, K.: Über Zusammenhänge zwischen dem CO_2-Austausch und der Abgabe von Wasserdampf durch *Bryophyllum daigremontianum* Berg. Planta 77, 212–223 (1967)
Kluge, M., Heininger, B.: Untersuchungen über den Efflux von Malat aus den Vacuolen der assimilierenden Zellen von *Bryophyllum daigremontianum* und mögliche Einflüsse dieses Vorganges auf den CAM. Planta *113*, 333–343 (1973)
Kluge, M., Kriebitsch, Ch., v. Willert, D.: Dark fixation of CO_2 in Crassulacean Acid Metabolism: Are two carboxylation steps involved? Z. Pflanzenphysiol. *72*, 460–467 (1974)
Kluge, M., Lange, O. L., von Eichmann, M., Schmidt, R.: Diurnaler Säurerhythmus bei *Tillandsia usneoides:* Untersuchungen über den Weg des Kohlenstoffs sowie die Abhängigkeit des CO_2-Gaswechsels von Lichtintensität, Temperatur und Wassergehalt der Pflanze. Planta *112*, 357–372 (1973)
Kluge, M., Lüttge, U.: The role of malic acid fluxes in regulation of Crassulacean Acid Metabolism (CAM): Osmoregulation of efflux at the tonoplast. In: Membrane Transport in Plants. Zimmermann, M., Dainty, J. (eds.), p. 101. Berlin, Heidelberg, New York: Springer 1974

Kluge, M., Osmond, C. B.: Pyruvate, Pi dikinase in Crassulacean Acid Metabolism. Naturwissenschaften *58*, 414–415 (1971)

Kluge, M., Osmond, C. B.: Studies on phosphoenolpyruvate carboxylase and other enzymes of crassulacean acid metabolism of *Bryophyllum tubiflorum* and *Sedum praealtum*. Z. Pflanzenphysiol. *66*, 97–105 (1972)

Knautt, R. L., Arditti, J.: Partial identification of dark CO_2 fixation products in leaves of *Cattleya (Orchidaceae)*. New Phytol. *68*, 657–661 (1969)

Kramer, D., Willert, D. J. von: Vacuole-like spaces in *Bryophyllum daigremontianum* and *Bryophyllum tubiflorum*. Naturwissenschaften *59*, 315–316 (1972)

Krapf, G., Jacobi, G.: Starch starvation and plant metabolism. III. CO_2 fixation in isolated chloroplasts. Planta *124*, 145–152 (1975)

Kraus, G.: Über die Wasserverteilung in den Pflanzen. IV. Die Acidität des Zellsaftes. Abh. Naturforsch. Ges. Halle *16*, 141–205 (1884)

Kristen, U.: Untersuchungen über den Zusammenhang zwischen dem CO_2-Gaswechsel und Luftwegigkeit an den CAM-Sukkulenten *Bryophyllum daigremontianum* Berg. und *Agave americana* L. Flora *160*, 127–138 (1969)

Krotkov, R., Runeckles, V. C., Thimann, K. V.: Effect of light on the CO_2 absorption and evolution by *Kalanchoe*, wheat and pea leaves. Plant Physiol. *33*, 289–292 (1958)

Kuiper, P. J. C.: The effects of environmental factors on the transpiration of leaves, with special references to stomatal light response. Meded Landbowhogesch. Wageningen *61*, 1–49 (1964)

Kunitake, G., Saltmann, P.: Dark fixation of CO_2 by succulent leaves: conservation of the dark fixed CO_2 under diurnal conditions. Plant Physiol. *83*, 400–403 (1958)

Laetsch, W. M.: Chloroplast specialication on Dicotyledons possessing the C_4-dicarboxylic acid pathway of photosynthetic CO_2 fixation. Am. J. Bot. *55*, 875–883 (1968)

Laetsch, W. M.: The C_4 syndrome: a structural analysis. Ann. Rev. Plant Physiol. *25*, 24–52 (1974)

Lai Hong-huei: The physiology of *Opuntia* and other CAM plants. Ph.D. Thesis, Univ. of Melbourne 1977

Lange, O. L., Lösch, R., Schulze, E. D., Kappen, L.: Responses of stomata to changes in humidity. Planta *100*, 76–86 (1971)

Lange, O. L., Schulze, E. D., Kappen, L., Evenari, M., Buschbom, U.: CO_2 exchange pattern under natural conditions of *Caralluma negevensis*, a CAM plant of the Negev desert. Photosynthetica *9*, 318–326 (1975)

Lange, O. L., Schwemmle, B.: Untersuchungen zur Hitzeresistenz vegetativer und blühender Pflanzen von *Kalanchoe blossfeldiana*. Planta *55*, 208–225 (1960)

Lange, O. L., Zuber, M.: *Frerea indica*, a stem succulent CAM plant with deciduous C_3 leaves. Oecologia *31*, 67–72 (1977)

Larcher, W.: Ökologie der Pflanzen. Stuttgart: Verlag Eugen Ulmer 1973

Larson, D. W., Kershaw, K. A.: Measurement of CO_2 exchange in lichens. A new method. Can. J. Bot. *53*, 1535–1541 (1975)

Lawrence, G. H. M.: Taxonomy of Vascular Plant. New York: The Macmillan Co. 1955

Lee, R. E., Thompson, A.: The stromacentre of plastids of *Kalanchoe pinnata* Persoon. J. Ultrastruct. Res. *42*, 451–456 (1973)

Leese, O., Leese, M.: Desert plants—cacti and succulents. W. H. Collingridge, L. Collingridge Limited, London: Florida Transatlantic Arts Incorporated 1959

LeHouerou, H. N.: North Africa: past, present, future. In: Arid lands in transition. Dregne, H. E. (ed.). Am. Assoc. Adv. Sci. Publ. *90*, 227–278 (1970)

Lerman, J. C.: How to interpret variations in the carbon isotope ratio of plants: biologic and environmental effects. In: Environmental and Biological Control of Photosynthesis. Marcelle, R. (ed.), pp. 323–336. The Hague: Dr. W. Junk b.v. Publishers 1975

Lerman, J. C., Deleens, E., Nato, A., Moyse, A.: Variation in the carbon isotope composition of a plant with Crassulacean Acid Metabolism. Plant Physiol. *53*, 581–584 (1974)

Levitt, J.: Physiological basis of stomatal response. In: Water and plant life. Ecological studies, Vol. 19. Lange, O. L., Kappen, L., Schulze, E. D. (eds.), pp. 160–168. Berlin, Heidelberg, New York: Springer 1976

Libbert, E.: Kompendium der allgemeinen Biologie. Stuttgart, New York: G. Fischer Verlag 1976

Lips, H. S., Beevers, H.: Compartmentation of organic acids in corn roots. I. Differential labeling of two malate pools. Plant Physiol. *41*, 709–712 (1966 a)

Lips, H. S., Beevers, H.: Compartmentation of organic acids in corn roots. II. The cytoplasmic pool of malic acid. Plant Physiol. *41*, 713–717 (1966 b)

Livingston, B. E.: Relative transpiration in cacti. Plant World *10*, 110–114 (1907)

Lowe, J., Slack, C. R.: Inhibition of maize leaf phosphoenolpyruvate carboxylase by oxaloacetate. Biochem. Biophys. Acta *235*, 207–209 (1971)

Ludlow, M. M.: Effect of oxygen concentration on leaf photosynthesis and resistances to carbon dioxide diffusion. Planta *91*, 285–290 (1970)

Ludlow, M. M., Wilson, G. L.: Photosynthesis of tropical plants. IV. Basis and consequences of differences between grasses and legumes. Aust. J. Bot. Sci. *25*, 1133–1145 (1972)

Lüttge, U., Ball, E.: Proton and malate fluxes in cells of *Bryophyllum daigremontianum* leaf slices in relation to potential osmotic pressure of the medium. Z. Pflanzenphysiol. *73*, 326–338 (1974)

Lüttge, U., Ball, E.: Water relation parameters of the CAM plant *Kalanchoe daigremontiana* in relation to diurnal malate oscillation. Oecologia *31*, 85–94 (1977)

Lüttge, U., Ball, E., Greenway, H.: Effects of water and turgor potential on malate efflux from leaf slices of *Kalanchoe daigremontiana*. Plant Physiol. *60*, 521–523 (1977)

Lüttge, U., Kluge, M., Ball, E.: Effects of osmotic gradients on vacuolar malic acid storage. A basic principle in oscillatory behaviour of Crassulacean Acid Metabolism. Plant Physiol. *56*, 613–616 (1975)

Lyndon, R. F.: Studies on nitrogen metabolism of *Kalanchoe* leaves in the dark in relation to acidification deacidification and gaseous exchange. Ph.D. Thesis, Durham Univ., Durham, Eng. (Newcastle Division) 1957

Lyndon, R. A.: Nitrogen metabolism of detached *Kalanchoe* leaves in the dark in relation to acidification, deacidification, and O_2 uptake. J. Exp. Bot. *13*, 20–35 (1962)

MacDougal, D. T.: A new high temperature record for growth. Science *53*, 370–372 (1921)

MacDougal, D. T., Working, E. B.: Another high temperature record for growth and endurance. Science *54*, 152–153 (1921)

MacLennan, D. H., Beevers, H., Harley, J. L.: "Compartmentation" of acids in plant tissues. Biochem. J. *89*, 316–327 (1963)

Madhusudana, Rao, I., Swamy, P. M., Das, V. S. R.: CAM syndrome in some nonsucculents and its inhibition by paraquot. Abstracts of 4th Int. Congress on Photosynthesis ISBN No. 0-9049 63-17-9, 1977

Maeba, P., Sanwal, B. D.: Feedback inhibition of phosphoenolpyruvate carboxylase in *Salmonella*. Biochem. Biophys. Res. Commun. *21*, 503–508 (1965)

Marcelle, R.: Effect of photoperiod on the CO_2 and O_2 exchanges in leaves of *Bryophyllum daigremontianum* (Berger). In: Environmental and biological control of photosynthesis. Marcelle, R. (ed.), pp. 349–356. The Hague: Dr. W. Junk b.v. Publishers, 1975

Martinez, M.: Plantas utiles de la flora mexicana. Cacteceas. Ediciones Botas Mexico 1959

Maruyama, H., Easterday, R. L., Chang, H. C., Lane, M. D.: The enzymatic carboxylation of phosphoenolpyruvate carboxylase. J. Biol. Chem. *241*, 2405–2412 (1966)

Mayer, A.: Die Sauerstoffausscheidung einiger dickblättriger Pflanzen bei Abwesenheit von Kohlensäure und die physiologische Bedeutung dieser Erscheinung. Landwirtschaftl. Vers. Stn. *34*, 127–143 (1887)

Mazelis, M., Vennesland, B.: Carbon dioxide fixation into oxaloacetate in higher plants. Plant Physiol. *32*, 591–600 (1957)

McLaren, I., Thomas, D. R.: CO_2 fixation, organic acids and some enzymes in green and colourless tissue cultures of *Kalanchoe crenata*. New Phytol. *66*, 683–695 (1967)

McWilliams, E. L.: Comparative rates of dark CO_2 uptake and acidification in *Bromeliaceae*, *Orchidaceae*, and *Euphorbiaceae*. Bot. Gaz. *131*, 285–290 (1970)

Medina, E.: Dark CO_2 fixation habitat preference and evolution within the *Bromeliaceae*. Evolution *28*, 677–686 (1974)

Medina, E., Delgado, M.: Photosynthesis and night CO_2 fixation in *Echeveria columbiana* v. Pellnitz. Photosynthetica *10*, 155–163 (1976)

Medina, E., Delgado, M., Troughton, J. H., Medina, J. D.: Physiological ecology of CO_2 fixation in *Bromeliaceae*. Flora *166*, 137–152 (1977)

Medina, E., Troughton, J. H.: Dark CO_2 fixation and the carbon isotope ratio in *Bromeliaceae*. Plant Sci. Lett. *2*, 357–362 (1974)

Meidner, H., Mansfield, T. A.: Physiology of stomata. London: McGraw-Hill 1968

Meinzer, F. C., Rundel, P. W.: Crassulacean Acid Metabolism and water use efficiency in *Echeveria pumila*. Photosynthetica *7*, 358–364 (1973)

Metcalfe, C. R., Chalk, L.: Anatomy of the Dicotyledons. Leaves, stem and wood relation to taxonomy with notes on economic uses, Vol. I, II. Oxford at the Clarendon Press 1950

Metral, J. J.: Les cactes fourragères dous le Nord-Est du Brasil plus particutirement dous l'état du ceara. Agron. Trop. *20*, 248–261 (1965)

Milburn, T. R., Pearson, D. J., Ndegwe, N. A.: Crassulacean Acid Metabolism under natural tropical conditions. New Phytol. *67*, 883–897 (1968)

Mohr, H.: Lehrbuch der Pflanzenphysiologie. Berlin, Heidelberg, New York: Springer 1969

Mooney, H., Troughton, J. H., Berry, J. A.: Arid climate and photosynthetic systems. Carnegie Inst. Yearb. *73*, 793–804 (1974)

Mooney, H. A., Troughton, J. H., Berry, J. A.: Carbon isotope ratio measurements of succulent plants of southern Africa. Oecologia *30*, 295–305 (1977a)

Mooney, H. A., Weisser, P. J., Gulmon, S. L.: Environmental adaptations of the Atacaman desert Cactus *Copiapoa haseltoniana*. Flora *166*, 117–124 (1977b)

Moradshahi, A., Vines, H., Black, J. C.: CO_2 exchange and acidity levels in detached pineapple [*Ananas comosus* (L.) Merr.], leaves during the day at various temperatures, O_2 and CO_2 concentrations. Plant Physiol. *59*(2), 274–278 (1977)

Morel, C., Celati, C., Queiroz, O.: Sur le métabolisme des Crassulacées. V. Adaptation de la capacité enzymatique aux changements de photoperiode (metabolisme de l'acide malique et transaminations). Physiol. Veg. *10*, 743–763 (1972)

Morel-Claret, C.: Photopériodisme et métabolisme acide d'une crassulacee *Kalanchoe blossfeldiana* "Tom Thumb" rhythmes circadiens d'activitiés enzymatiques et regulation. Ph.D. Thesis, Univ. of Paris 1973

Moyse, A.: Le métabolisme des acides organiques chez le *Bryophyllum* (Crassulacée). I. Oxydations respiratoires et fixation de l'anhydride carbonique à l'obscurité par β-carboxylation, en function de la tension d'oxygène. Physiol. Plant. *8*, 453–477 (1955a)

Moyse, A.: Le metabolisme des acides organiques chez *Bryophyllum* (Crassulacee). II. Les variations de l'acidité et la photosynthèse, en fonction de la tension d'oxygène. Physiol. Plant. *8*, 478–492 (1955b)

Moyse, A., Jolchine, G.: Les variations quantitatives des acides organiques des feuilles de *Bryophyllum*, à l'obscurité et a la lumière, en function de la tension partielle de l'oxygène. Bull. Soc. Chim. Biol. *39*, 725–784 (1956)

Moyse, A., Jolchine, G.: L-action de la lumière sur la β-carboxylation et les oxydations dans les feuilles de *Bryophyllum*. Bull. Soc. Chim. Biol. *39*, 725–744 (1957)

Mukerji, S. K.: Four-hourly variations in the activities of malate dehydrogenase (decarboxylating) and phosphopyruvate carboxylase in the cactus (*Nopalea dejecta*) plant. Indian J. Biochem. *5*, 62–64 (1968)

Mukerji, S. K.: Corn leaf phosphoenolpyruvate carboxylases: activation by magnesium ions. Plant Sci. Lett. *2*, 243–248 (1974)

Mukerji, S. K., Ting, I. P.: Malate dehydrogenase (decarboxylating) (NADP) isoenzymes of *Opuntia* stem tissue. Mitochondria, chloroplast, and soluble forms. Biochem. Biophys. Acta *167*, 239–249 (1968a)

Mukerji, S. K., Ting, I. P.: Intracellular localization of CO_2 metabolism enzymes in cactus phylloclades. Phytochemistry *7*, 903–911 (1968b)

Mukerji, S. K., Ting, I. P.: Malic dehydrogenase isoenzymes in green stem tissue of *Opuntia*: isolation and characterization. Arch. Biochem. Biophys. *131*, 336–351 (1969)

Mukerji, S. K., Ting, I. P.: Phosphoenolpyruvate carboxylase isoenzymes: separation and properties of three forms from cotton leaf tissue. Arch. Biochem. Biophys. *143*, 297–317 (1971)

Nalborczyk, E., LaCroix, L. J., Hill, R. D.: Environmental influences on light and dark CO_2 fixation by *Kalanchoe daigremontiana*. Can. J. Bot. *53*, 1132–1138 (1975)

Neales, T. F.: Effect of ambient carbon dioxide concentration on the rate of transpiration of *Agave americana* in the dark. Nature (London) *228*, 880–882 (1970)

Neales, T.F.: The effect of night temperature and CO_2 assimilation transpiration and water use efficiency in *Agave americana* L. Aust. J. Biol. Sci. *26*, 705–714 (1973a)
Neales, T.F.: Effect of night temperature on the assimilation of carbon dioxide by mature pineapple plants *Ananas comosus* (L.) Merr. Aust. J. Biol. Sci. *26*, 539–546 (1973b)
Neales, T.F.: The gas exchange pattern of CAM plants. In: Environmental and biological control of photosynthesis. Marcelle, R. (ed.), pp. 299–310. The Hague: Dr. W. Junk b.v. Publishers, 1975a
Neales, T.F., Hew, C.S.: Two types of carbon fixation in tropical orchids. Planta *123*, 303–306 (1975)
Neales, T.F., Patterson, A.A., Hartney, V.J.: Physiological adaptation to growth in the carbon assimilation and water loss of xerophytes. Nature (London) *219*, 469–472 (1968)
Nishida, K.: Studies on stomatal movement of crassulacean plants in relation to the acid metabolism. Physiol. Plant. *16*, 281–298 (1963)
Nishida, K.: CO_2 fixation in leaves of a CAM plant without lower epidermis and the effect of CO_2 on their deacidification. Plant Cell Physiol. *18*, 927–930 (1977)
Nobel, P.S.: Boundary layers of air adjacent to cylinders. Estimation of effective thickness and measurements on plant material. Plant Physiol. *54*, 177–181 (1974)
Nobel, P.S.: Effective thickness and resistance of the air boundary layer adjacent to spherical plant parts. J. Exp. Bot. *26*, 120–130 (1975)
Nobel, P.S.: Water relations and photosynthesis of a desert CAM plant, *Agave deserti*. Plant Physiol. *58*, 576–582 (1976)
Nobel, P.S.: Water relations of flowering of *Agave deserti*. Bot. Gaz. *138*, 1–6 (1977a)
Nobel, P.S.: Water relations and photosynthesis of a Barrel cactus, *Ferocactus acanthodes*, in the colorado desert. Oecologia *27*, 117–133 (1977b)
Nordal, A., Benson, A.A.: Phorbic acid biosynthesis in the latex vessel system of *Euphorbia*. Plant Physiol. *44*, 78–84 (1969)
Nordal, A., Ogner, G.: The detection of phorbic acid in *Euphorbia palustris* L. Acta Chem. Scand. *18*, 830 (1964)
Nose, A., Shiroma, M., Miyazato, K., Murayama, S.: Studies on matter production in pineapple plants. I. Effects of light intensity in light period on CO_2 exchange and CO_2 balance of pineapple plants. Jap. J. Crop Sci. *46*, 579–587 (1977)
Nuernbergk, E.L.: Über den zeitlichen Verlauf der Photosynthese bei Gewächshauspflanzen. Gartenbauwissenschaft *19*, 391–398 (1955)
Nuernbergk, E.L.: Weitere Beiträge zum Kohlendioxyd-Stoffwechsel von Pflanzen mit diurnalem Säurerhythmus von Lang- und Kurztagspflanzen. Mitt. Inst. Allg. Bot. Hamburg *11*, 205–232 (1957)
Nuernbergk, E.L.: Endogener Rhythmus und CO_2 Stoffwechsel bei Pflanzen mit diurnalem Säurerhythmus. Planta *56*, 28–70 (1961)
Nuernbergk, E.L.: Temperatur und Kohlendioxid-Stoffwechsel bei *Bryophyllum daigremontianum*. Port. Acta Biol. *6*, 298–358 (1962)
Ogren, W.L.: Control of photorespiration in soybean and maize. In: Environmental and biological control of photosynthesis. Marcelle, R. (ed.), pp. 45–52. The Hague: Dr. W. Junk b.v. Publishers, 1975
Osmond, C.B.: β-carboxylation during photosynthesis in Atriplex. Biochim. Biophys. Acta *141*, 197–199 (1967)
Osmond, C.B.: Environmental control of photosynthetic options in crassulacean plants. In: Environmental and biological control of photosynthesis. Marcelle, R. (ed.), pp. 311–321. The Hague: Dr. W. Junk b.v. Publishers, 1975
Osmond, C.B.: CO_2 assimilation and dissimilation in the light and dark in CAM plants. In: CO_2 metabolism and plant productivity. Burris, R.H., Black, C.C. (eds.), pp. 217–233. Baltimore, London, Tokyo: University Park Press 1976
Osmond, C.B.: Ion absorption and carbon metabolism in cells of higher plants. In: Transport in plants II, Part A. Enzyclopedia of plant physiology. New Series, Vol. II. Lüttge, U., Pitman, M G. (eds.), pp. 347–366. Berlin, Heidelberg, New York: Springer 1976b
Osmond, C.B.: Crassulacean Acid Metabolism: A curiosity in context. Ann. Rev. Plant Physiol. *29*, 379–414 (1978)

Osmond, C. B., Allaway, W. G.: Pathway of CO_2 fixation in the CAM plant *Kalanchoe daigremontiana*. I. Patterns of $^{14}CO_2$ fixation in the light. Aust. J. Plant Physiol. *1*, 503–512 (1974)

Osmond, C. B., Allaway, W. G., Sutton, B. G., Troughton, J. H., Queiroz, O., Lüttge, U., Winter, K.: Carbon isotope discrimination in photosynthesis of CAM plants. Nature (London) *246*, 41–42 (1973)

Osmond, C. B., Avadhani, P. N.: Acid metabolism in *Atriplex*. II. Oxalate synthesis during acid metabolism in the dark. Aust. J. Biol. Sci. *21*, 917–927 (1968)

Osmond, C. B., Bender, M. M., Burris, R. H.: Pathways of CO_2 fixation in the CAM plant *Kalanchoe daigremontiana*. III. Correlation with $\delta^{13}C$ value during growth and water stress. Aust. J. Plant Physiol. *3*, 787–799 (1976)

Osmond, C. B., Björkman, O.: Pathways of CO_2 fixation in the CAM plant *Kalanchoe daigremontiana*. II. Effects of O_2 and CO_2 concentration on light and dark CO_2 fixation. Aust. J. Plant Physiol. *2*, 155–162 (1975)

Osmond, C. B., Greenway, H.: Salt responses of carboxylation enzymes from species differing in salt tolerance. Plant Physiol. *49*, 260–262 (1972)

Osmond, C. B., Laties, G. G.: Compartmentation of malate in relation ion absorption in beet. Plant Physiol. *44*, 7–14 (1969)

Osmond, C. B., Nott, D. L., Zinth, P. M.: Carbon assimilation patterns and growth of the introduced CAM plant *Opuntia inermis* in Eastern Australia. Oecologia, in press (1978)

Osmond, C. B., Ziegler, H., Stichler, W., Trimborn, P.: Carbon isotope discrimination in alpine succulent plants supposed to be capable of Crassulacean Acid Metabolism (CAM). Oecologia *18*, 209–217 (1975)

Pallas, J. E., Wright, B. G.: Organic acid changes in the epidermis of *Vicia faba* and their implication in stomatal movements. Plant Physiol. *51*, 588–590 (1973)

Paredes-Lopez, O. G., Vital-Bari, Canargo-Rubio, E.: The influence of dissolved oxygen in biomass production from nopal fruit. J. Ferm. Tech. *52*, 592–597 (1974)

Patten, D. T., Dinger, B. E.: Carbon dioxide exchange patterns of cacti from different environments. Ecology *50*, 686–688 (1969)

Peak, M. J., Peak, J. G., Ting, I. P.: Function and regulation of NADP specific malate dehydrogenase in *Euglena gracilis*. Z. Biochim. Biophys. Acta *293*, 312–321 (1973)

Penman, H. L., Schofield, R. K.: Some physical aspects of assimilation and transpiration. Symp. Soc. Exp. Biol. *5*, 115–129 (1951)

Pfeffer, W.: Pflanzenphysiologie. Ein Handbuch der Lehre vom Stoffwechsel und Kraftwechsel in der Pflanze, Vol. 1, Stoffwechsel, pp. 310. Leipzig: Verlag Wilhelm Engelmann 1897

Pieper, R. D.: Blue grama vegetation responds inconsistently to cholla cactus control. J. Range Manage. *24*, 52–54 (1971)

Pina, L. I.: Algunas species del genero Yucco del nareste de Mexico. Cact. Suculentas Mex. *24*, 3–11 (1974)

Pittendrigh, C. S.: On temperature independence in the clock system controlling emergence tissue in *Drosophila*. Proc. Natl. Acad. Sci. U.S.A. *40*, 1018–1029 (1954)

Powell, R. W.: Further experiments on the evaporation of water from saturated surfaces. Trans. Inst. Chem. Eng. *13*, 175–198 (1940)

Pucher, G. W., Vickery, H. B., Abrahams, M. D., Leavenworth, C. S.: Studies in the metabolism of crassulacean plants: Diurnal variation in organic acids and starch in excised plants *Bryophyllum calycinum*. Plant Physiol. *24*, 610–620 (1949)

Queiroz, O.: Sur le métabolisme acide des Crassulacées. I. Action à long terme de la température de nuit sur la synthèse d'acide malique par *Kalanchoe blossfeldiana* "Tom Thumb" placée en jours courts. Physiol. V. *3*, 203–213 (1965)

Queiroz, O.: Sur le métabolisme acide des Crassulaceées. II. Action à long terme de la température de jour sur les variations de la teneur en acide malique en jours courts. Physiol. V. *4*, 323–339 (1966)

Queiroz, O.: Recherche d'un modèle enzymatique pour le déterminisme de la désacidification diurne chez les Crassulacées. C.R. Acad. Sci. *265*, 1928–1931 (1967)

Queiroz, O.: Sur le métabolisme acide des Crassulacées. III. Variations d'activité enzymatique sous l'action du photopériodisme et du thermopériodisme. Physiol. V. *6*, 117–136 (1968a)

Queiroz, O.: Action du photopériodisme sur l'activité enzymatique dans la synthèse au la dégradation de l'acide malique chez *Kalanchoe blossfeldiana*. C.R. Acad. Sci. *266*, 1260–1262 (1968 b)
Queiroz, O.: Sur le métabolisme acide des Crassulacées. IV. Réflexions sur les phenomenes oscillatoires au niveau enzymatique et sur la compartimentation métabolique, sous l'action du photopériodisme. Physiol. Veg. *8*, 75–110 (1970)
Queiroz, O.: Circadian rhythms in enzyme activities and photochrome. In: Phytochrome. Shropshire, W., Mitrakos, K. (eds.), pp. 295–316. New York: Academic Press 1972
Queiroz, O.: Circadian rhythms and metabolic patterns. Ann. Rev. Plant Physiol. *24*, 115–134 (1974)
Queiroz, O.: Rhythmical characteristics of different levels of CAM regulation. Physiological and adaptive significance. In: Environmental and biological control of photosynthesis. Marcelle, R. (ed.), pp. 357–368. The Hague: Dr. W. Junk b.v. Publishers, 1975
Queiroz, O., Celati, C., Morel, C.: Sur le métabolisme acid des Crassulacées. VI. Photopériodisme et rhythmes circadiens d'activité enzymatique. Systèmes régulateurs. Physiol. Veg. *10*, 765–781 (1972)
Queiroz, O., Morel, C.: Photoperiodism and enzyme activity: towards a model for the control of circadian metabolic rhythms in the crassulacean acid metabolism. Plant Physiol. 596–602 (1974)
Queiroz, O., Morel, C., Celati, C.: Photopériodisme et activité enzymatique: adaptation à long terme, d'un réseau métabolique complexe, aux changements de photoperiode. C.R. Acad. Sci. *272*, 3045–3048 (1971)
Queiroz, O., Trippi, V. S.: Physiologie végétale—action du photopériodisme sur l'activité enzymatique: les isozymes de la malique déshydrogénase. C.R. Acad. Sci. *268*, 2060–2062 (1969)
Ranson, S. L., Thomas, M.: Crassulacean Acid Metabolism. Ann. Rev. Plant Physiol. *11*, 81–110 (1960)
Raschke, K.: Die Reaktionen des CO_2-Regelsystems in den Schließzellen von *Zea mays* auf weißes Licht. Planta *68*, 111–140 (1966)
Raschke, K.: Stomatal action. Ann. Rev. Plant Physiol. *26*, 309–340 (1975 a)
Raschke, K.: Simultaneous requirement of carbon dioxide and abscisic acid for stomatal closing in *Xanthium strumarium* L. Planta *125*, 243–259 (1975 b)
Raschke, K.: How stomata resolve the dilemma of opposing priorities. Philos. Trans. R. Soc. London. B 273 (927), 551–560 (1976)
Richards, H. M.: Acidity and gas exchange in cacti. Carnegie Inst. Washington Publ. *208*, 107 (1915)
Rocha, V., Ting, I. P.: Tissue distribution of microbody, mitochondrial, and soluble malate dehydrogenase isoenzymes. Plant Physiol. *46*, 754–756 (1970)
Rosenberg, L. L., Capindale, J. B., Whatley, F. R.: Formation of oxaloacetate and aspartate from phosphoenol-pyruvate in spinach leaf chloroplast extract. Nature (London) *181*, 632–633 (1958)
Rouhani, I.: Pathways of carbon metabolism in spongy mesophyll cells isolated from *Sedum telephium* leaves and their relationship to Crassulacean Acid Metabolism plants. Ph.D. Thesis, Univ. Georgia, Athens 1972
Rouhani, I., Vines, H. M., Black, C. C., Jr.: Isolation of mesophyll cells from *Sedum telephium* leaves. Plant Physiol. *51*, 97–103 (1973)
Saltman, P., Kunitake, G., Spolter, H., Stitt, C.: The dark fixation of CO_2 by succulent leaves: the first products. Plant Physiol. *31*, 464–468 (1956)
Saltman, P., Lynch, V. H., Kunitake, G. M., Stitt, C., Spolter, H.: The dark fixation of CO_2 by succulent leaves, metabolic changes subsequent to initial fixation. Plant Physiol. *32*, 197–200 (1957)
Sanwal, B. D.: Regulatory mechanisms involving nicotinamide adenine nucleotides as allosteric effectors. I. Control characteristics of malate dehydrogenase. J. Biol. Chem. *244*, 1831–1837 (1969)
Schmitz, J.: Über Beziehungen zwischen Blütenbildung in verschiedenen Licht-Dunkel-Kombinationen und Atmungsrhythmik bei wechselnden photoperiodischen Bedingungen (Untersuchungen der *Kalanchoe blossfeldiana*). Planta *39*, 271–308 (1951)

Schulze, E.-D., Lange, O. L., Evenari, M., Kappen, L., Buschbom, U.: The role of air humiditiy and leaf temperature in controlling stomatal resistance of Prunus armeniaca L. under desert conditions. I. A simulation of the daily course of stomatal resistance. Oecologia *17*, 159–170 (1974)
Schulze, E.-D., Schulze, I.: Distribution and control of photosynthetic pathways in plants growing in the Namib Desert, with special regard to *Welwitschia mirabilis* Hook. fil. Madoqua, Nr. 3. 1976, Vol. 9, 5–13
Schulze, E.-D., Ziegler, H., Stichler, W.: Environmental control of Crassulacean Acid Metabolism in *Welwitschia mirabilis* Hook. fil. in its range of natural distribution in the Namib desert. Oecologia *24*, 323–334 (1976)
Schütte, K. H., Steyn, R., Westhuizen, Van der M.: Crassulacean Acid Metabolism in South African succulents: a preliminary investigation into its occurence in various families. J. Soc. Afr. Bot. *33*, 107–110 (1967)
Schwemmle, B., Lange, O. L.: Endogen-tagesperiodische Schwankungen der Hitzeresistenz bei *Kalanchoe blossfeldiana*. Planta *53*, 134–144 (1959)
Sen, D. N.: A report of Crassulacean Acid Metabolism in *Euphorbia caducifolia* and its diversity. Curr. Sci. *39*, 116–117 (1970)
Seshagiri, P. V. V., Suryanaryana-Murthy, A.: Studies in the organic acid metabolism of *Ananas sativa* L. J. Indian Bot. Soc. *36*, 254–261 (1951)
Shreve, E. B.: The daily march of transpiration in a desert perenial. Carnegie Inst. Washington Publ. *194*, 64 (1914)
Shreve, E. B.: An analysis of the causes of variations in the transpiring power of cacti. Physiol. Res. *2*, 73–127 (1916)
Sideris, C. P., Krauss, G. H.: Transpiration and translocation phenomena in pine apple. Am. J. Bot. *42*, 707–709 (1955)
Slack, C. R., Hatch, M. D., Goodchild, D. J.: Distribution of enzymes in mesophyll and parenchyma-sheath chloroplasts of maize leaves in relation to the C_4-dicarboxylic acid pathway of photosynthesis. Biochem. J. *114*, 489–498 (1969)
Slatyer, R. O.: Plant-water relationships. London, New York: Academic Press 1967
Smith, B. N., Epstein, S.: Two categories of $^{13}C/^{12}C$ ratios for higher plants. Plant Physiol. *47*, 380–384 (1971)
Smith, F. A., Raven, J. A.: H^+-Transport and regulation of cell pH. In: Transport in plants II, Part A. Enzyclopedia of plant physiology, Vol. II. Lüttge, U., Pitman, M. G. (eds.), pp. 317–346. Berlin, Heidelberg, New York: Springer 1976
Smith, T. E.: Partial purification and characteristics of potato phosphoenolpyruvate carboxylase. Arch. Biochem. Biophys. *125*, 178–188 (1968)
Somers, G. F.: The influence of light, temperature, and some enzyme poisons upon the total organic aci= content of leaf tissue of *Kalanchoe daigremontiana* Hamet et Perrier. Plant Physiol. *26*, 1–18 (1951)
Spear, J., Thimann, K.: The interrelation between CO_2 metabolism and photoperiodism in *Kalanchoe*. II. Effect of prolonged darkness and high temperatures. Plant Physiol. *29*, 414 (1954)
Sphoer, H. A.: The carbohydrate economy of cacti. Carnegie Inst. Washington Publ. *287*, 79 (1919)
Stiller, M. L.: The mechanism of malate synthesis in *Crassulacean* leaves. Ph.D. Thesis, Purdue Univ., Lafayette, Ind. 1959
Stocker, O.: Grundriß der Botanik. Berlin, Göttingen, Heidelberg: Springer 1952
Stocker, O.: Die Abhängigkeit der Transpiration an den Umweltfaktoren. In: Handbuch der Pflanzenphysiologie, Vol. V. Ruhland (ed.), pp. 436–488. Berlin, Göttingen, Heidelberg: Springer 1956
Stocker, O.: Die photosynthetischen Leistungen der Steppen- und Wüstenpflanzen. In: Handbuch der Pflanzenphysiol., Vol. V. Ruhland (ed.), pp. 460–491. Berlin, Göttingen, Heidelberg: Springer 1960
Sugiyama, T., Laetsch, W. M.: Occurence of pyruvate orthophosphate dikinase in the succulent plant, *Kalanchoe daigremontiana* Hamet et Perr. Plant Physiol. *56*, 605–607 (1975)
Sutton, B. C.: Regulation of carbohydrate metabolism in succulent plants. Ph.D. Thesis, Aust. Nat. Univ. (1974a)

Sutton, B.G.: The regulation of carbohydrate metabolism in Crassulacean plants at night. Plant Physiol. Ann. *54* (1974b)

Sutton, B.G.: Glycolysis in CAM plants. Aust. J. Plant Physiol. *2*, 389–402 (1975a)

Sutton, B.G.: The path of carbon in CAM plants at night. Aust. J. Plant Physiol. *2*, 377–388 (1975b)

Sutton, B.G.: Control of glycolysis in succulent plants at night. In: Environmental and biological control of photosynthesis. Marcelle, R. (ed.), pp. 337–348. The Hague: Dr. W. Junk b.v. Publishers, 1975c

Sutton, B.G., Osmond, C.B.: Dark fixation of CO_2 by *Crassulacean* plants. Evidence for a single carboxylation step. Plant Physiol. *50*, 360–365 (1972)

Sutton, B.G., Ting, I.P., Troughton, J.H.: Seasonal effects on carbon isotope composition of cactus in a desert environment. Nature (London) *261*, 42–43 (1976)

Sweeny, B.M.: Rhythmic phenomena in plants. London, New York: Academic Press 1969

Szarek, S.R.: Physiological mechanisms of drought adaptation in *Opuntia basilaris* Engelm. Ph.D. Thesis, Univ. Calif. Riverside 1974

Szarek, S.R., Johnson, H.B., Ting, I.P.: Drought adaptation of *Opuntia basilaris*. Significance of recycling carbon through Crassulacean Acid Metabolism. Plant Physiol. *52*, 539–541 (1973)

Szarek, S.R., Ting, I.P.: Respiration and gas exchange in stem tissue of *Opuntia basilaris*. Plant Physiol. *54*, 829–834 (1974a)

Szarek, S.R., Ting, I.P.: Seasonal patterns of acid metabolism and gas exchange in *Opuntia basilaris*. Plant Physiol. *54*, 76–81 (1974b)

Szarek, S.R., Ting, I.P.: Photosynthetic efficiency of CAM plants in relation to C_3 and C_4 plants. In: Environmental and biological control of photosynthesis. Marcelle, R. (ed.), pp. 289–298. The Hague: Dr. W. Junk b.v. Publishers, 1975a

Szarek, S.R., Ting, I.P.: Physiological responses to rainfall in *Opuntia basilaris*. Am. J. Bot. *62*, 602–609 (1975b)

Szarek, S.R., Ting, I.P.: The occurrence of Crassulacean Acid Metabolism among plants. Photosynthetica *11*, 330–342 (1977)

Szarek, S.R., Troughton, J.H.: Carbon isotope ratios in Crassulacean Acid Metabolism plants. Seasonal patterns from plants in natural stands. Plant Physiol. *58*, 367–370 (1976)

Takhtajan, A.: Flowering plants. Origin and dispersal. Washington: Smithsonian Institution Press 1969

Thoday, D., Evans, H.: Buffer Systems in *Kleinia articulata*. Protoplasma *14*, 64–74 (1931)

Thomas, M.: Physiological studies on acid metabolism on green plants. I. CO_2 fixation and CO_2 liberation in Crassulacean Acid Metabolism. New Phytol. *48*, 390–420 (1949)

Thomas, M.: Carbon dioxide fixation and acid synthesis in Crassulacean Acid Metabolism. Symp. Soc. Exp. Biol. *5*, 72–93 (1951)

Thomas, M., Beevers, H.: Physiological studies on acid metabolism in green plants. II. Evidence of CO_2 fixation in *Bryophyllum* and the study of diurnal variation of acidity in this genus. New Phytol. *48*, 421–447 (1949)

Thomas, M., Ranson, S.L.: Physiological studies on acid metabolism in green plants. III. Further evidence of CO_2 fixation during dark acidification of plants showing Crassulacean Acid Metabolism. New Phytol. *53*, 1–30 (1954)

Thomson, W.W., Journett, de R.: Studies on the ultrastructure of the guard cells of *Opuntia*. Am. J. Bot. *57*, 309–316 (1970a)

Thomson, W.W., Journett, de R.: Observations on inclusion bodies in subsidary cells of *Opuntia* sp. J. Ultrastruct. Res. *31*, 439–443 (1970b)

Thomson, W.W., Platt, K.: Plastid ultrastructure in the barrel cactus, *Echinocactus acanthodes*. New Phytol. *72*, 791–797 (1973)

Thompson, A., Vogel, J., Lee, R.: Carbon dioxide uptake in relation to a plastid inclusion body in the succulent *Kalanchoe pinnata* Persoon. J. Exp. Bot. *28*, 1037–1041 (1977)

Thurlow, J., Bonner, J.: Fixation of atmospheric CO_2 in the dark by leaves of *Bryophyllum*. *Arch. Biochem. 19*, 509–511 (1948)

Ting, I.P.: CO_2 metabolism in corn roots. III. Inhibition of P-enolpyruvate carboxylase by L-malate. Plant Physiol. *43*, 1919–1924 (1968a)

Ting, I.P.: Malic dehydrogenase in corn root tips. Arch. Biochem. Biophys. *126*, 1–7 (1968b)

Ting, I. P.: Nonautotrophic CO_2 fixation and crassulacean acid metabolism. In: Photosynthesis and photorespiration. Hatch, M. D., Osmond, C. B. (eds.), pp. 169–185. New York: Wiley-Interscience 1971

Ting, I. P.: Malate dehydrogenase and other enzymes of C_4 acid metabolism in marine plants. Aust. J. Plant Physiol. *3*, 121–127 (1976a)

Ting, I. P.: Crassulacean acid metabolism in natural ecosystems in relation to annual CO_2 uptake pattern and water utilization. In: CO_2 metabolism and plant productivity. Burris, R. H., Black, C. C. (eds.), pp. 251–268. Baltimore, London, Tokyo: Univ. Park Press 1976b

Ting, I. P., Dean, M. L., Dugger, W. M. Jr.: Leaf resistance in succulent plants. Nature (London) *213*, 526–527 (1967)

Ting, I. P., Dugger, W. M.: Separation and detection of organic acids on silicia gel. Anal. Biochem. *12*, 571–578 (1965)

Ting, I. P., Dugger, W. M.: CO_2 fixation in *Opuntia* roots. Plant Physiol. *41*, 500–505 (1966)

Ting, I. P., Dugger, W. M.: CO_2 metabolism in corn roots. I. Kinetics of carboxylation and decarboxylation. Plant Physiol. *42*, 712–718 (1967)

Ting, I. P., Dugger, W. M.: Non-autotrophic carbon dioxide metabolism in cacti. Bot. Gaz. *129*, 9–15 (1968)

Ting, I. P., Hanscom, Z.: Induction of acid metabolism in *Portulacaria* afra. Plant Physiol. *59*, 511–514 (1977)

Ting, I. P., Jennings, W.: Deep Canyon, a Desert Wilderness for Science. The Philip L. Boyd Deep Canyon Desert Research Center, Palm Desert. University of California (1976)

Ting, I. P., Johnson, H. B., Szarek, S. R.: Net CO_2 fixation in crassulacean acid metabolism plants. In: Net Carbon Dioxide Assimilation in Higher Plants. Black, C. C. (ed.), pp. 26–53. Proc. Symp. S. Sect. Am. Soc. Plant Physiol. Cotton Inc. Raleigh 1972

Ting, I. P., Johnson, H. B., Szarek, S. R., Brum, G. D.: Gas Exchange and Productivity for Opuntia ssp. Riverside Prog. Rep. 1971

Ting, I. P., Osmond, C. B.: Photosynthetic phosphoenolpyruvate carboxylase. Characteristics of alloenzymes from leaves of C_3 and C_4 plants. Plant Physiol. *51*, 439–447 (1973a)

Ting, I. P., Osmond, C. B.: Multiple forms of plant phosphoenolpyruvate carboxylase associated with different metabolic pathways. Plant Physiol. *51*, 448–453 (1973b)

Ting, I. P., Osmond, C. B.: Activation of plant P-enolpyruvate carboxylases by glucose-6-phosphate. A particular role in Crassulacean Acid Metabolism. Plant Sci. Lett. *1*, 123–128 (1973c)

Ting, I. P., Rocha, V.: NADP specific malate dehydrogenase of green spinach leaf tissue. Arch. Biochem. Biophys. *147*, 156–164 (1971)

Ting, I. P., Szarek, S. R.: Drought adaptation in Crassulacean Acid Metabolism plants. In: Environmental physiology of desert organisms. Hadley, N. F. (ed.), pp. *152–167*. Strandsburg: Dowden, Hutchinson and Rass 1975

Ting, I. P., Thompson, M. L., Dugger, W. M. Jr.: Leaf resistance to water vapor transfer in succulent plants: effect of thermoperiod. Am. J. Bot. *54*, 245–251 (1967)

Ting, I. P., Zschoche, W. C.: Asparagine biosynthesis by cotton roots. Plant Physiol. *45*, 429–434 (1970)

Tolbert, N. E.: Microbodies-Peroxisomes and glyoxysomes. Ann. Rev. Plant Physiol. *22*, 45–74 (1971)

Tregunna, E. B., Krotkov, G., Nelson, C. D.: Effect of light intensity during illumination on carbon dioxide evolution in the following dark period. Plant Physiol. *36*, 31 (1961)

Treichel, S.: The effect of NaCl on the concentration of proline in different halophytes. Z. Pflanzenphysiol. *76*, 56–68 (1975)

Treichel, S., Bauer, P.: Unterschiedliche NaCl-Abhängigkeit des tagesperiodischen CO_2-Gaswechsels bei einigen halisch wachsenden Pflanzen. Oecologia *17*, 87–95 (1974)

Treichel, S., Kirst, G. O., Willert, D. J. von: Veränderung der Aktivität der Phosphoenolpyruvat-carboxylase durch NaCl bei Halophyten verschiedener Biotope. Z. Pflanzenphysiol. *71*, 437–449 (1974)

Troughton, J. H., Card, K. A.: Temperature effects on the carbon isotope ratio of C_3, C_4, and Crassulacean Acid Metabolism (CAM) plants. Planta *123*, 185–190 (1975)

Troughton, J. H., Card, K. A., Björkmann, O.: Temperature effects on the carbon isotope ratio of C_3, C_4, and CAM plants. Carnegie Inst. Yearb. *73*, 780–783 (1974)
Troughton, J. H., Mooney, H. A., Berry, J. H., Verity, D.: Variable carbon isotope ratios of *Dudlya* species growing in natural environments. Oecologia *30*, 307–311 (1977)
Varner, J. E., Burrell, R. C.: Use of $^{14}CO_2$ in the study of the acid metabolism of *Bryophyllum calycinum*. Arch. Biochem. *25*, 280–287 (1950)
Verduin, J.: Diffusion through multiperforated septa. In: Photosynthesis in plants. Franck, J., Loomis, W. E. (eds.), pp. 95–112. Ames, Iowa: Iowa State College Press 1949
Vickery, H. B.: The behaviour of isocitric acid in excised leaves of *Bryophyllum calycinum* during culture in alternating light and darkness. Plant Physiol. *27*, 9–17 (1952)
Vickery, H. B.: The behaviour of the organic acids and starch of *Bryophyllum calycinum* during culture in continous light. J. Biol. Chem. *205*, 369–381 (1953)
Vickery, H. B.: The effect of temperature on the behaviour of malic acid and starch in leaves of *Bryophyllum calycinum* cultured in darkness. Plant Physiol. *29*, 385–392 (1954a)
Vickery, H. B.: The effect of abnormally prolonged alternating periods of light and darkneess upon the composition of *Bryophyllum calycinum*. Plant Physiol. *29*, 520–526 (1954b)
Vickery, H. B.: The information of starch and the behaviour of iso-citric acid in excised leaves of *Bryophyllum* calycinum cultured in darkness. Plant Physiol. *32*, 220–224 (1957)
Vieweg, G. H., Fekete, de M. A. R.: Tagesgang der Amylaseaktivität im Blatt von *Kalanchoe daigremontiana*. Z. Pflanzenphysiol. *81*, 74–79 (1977)
Volk, O. H.: IV. Geographical distribution. In: A handbook of succulent plants. Jacobsen, H. (ed.), pp. 928–930. London: Blandford Press 1960
Walker, D. A.: Enzymatically active extracts from crassulacean plants and their relationship to dark acidification. Ph.D. Thesis, Durham Univ., Durham Eng. (Newcastle Division) 1956a
Walker, D. A.: Malate synthesis in a cell free extract from a crassulacean plant. Nature (London) *178*, 593–594 (1956b)
Walker, D. A.: Physiological studies on acid metabolism in green plants. Biochem. J. *67*, 73–79 (1957)
Walker, D. A.: Physiological studies on acid metabolism. VII. Malic enzyme from *Kalanchoe crenata:* effects of carbon dioxide concentration. Biochem. J. *74*, 216–223 (1960)
Walker, D. A.: Pyruvate carboxylation in plant metabolism. Biol. Rev. *37s; 215–256 (1962)*
Walter, H.: Die Anpassungen der Pflanzen an Wassermangel: Das Xerophytenproblem in kausal-physiologischer Betrachtung. Naturwissenschaft und Landwirtschaft, Heft 9. Freising-München: Verlag Dr. F. P. Da Herer u. Cie. 1926
Warren, D. M., Wilkins, M. B.: An endogenous rhythm in the rate of dark fixation of carbon dioxide in leaves of *Bryophyllum fedtschenkoi*. Nature (London) *191*, 686–688 (1961)
Waygood, E. R., Mache, R., Tan, C. R.: Carbon dioxide, the substrate for phosphoenolpyruvate carboxylase from leaves of maize. Can. J. Bot. *47*, 1455–1458 (1969)
Webb, K. L., Burley, J. W. A.: Dark fixation of $^{14}CO_2$ by obligate and facultative salt marsh halophytes. Can. J. Bot. *43*, 281–285 (1965)
Wee, Y. C.: Planting density trials with *Ananas comosus* (L.). Merr. Var. Singapore Spanisj, Malay. Agric. J. *47*, 164–174 (1969)
Weinberg, R.: Effect of sodium chloride on the activity of a soluble malate dehydrogenase from pea seeds. J. Biol. Chem. *242*, 3000–3006 (1967)
Whelan, T., Sackett, W. M., Benedict, C. R.: Enzymatic fractionating of carbon isotopes by phosphoenolpyruvate carboxylase from C_4 plants. Plant Physiol. *51*, 1051–1054 (1973)
Whelan, T. W., Sackett, N., Benedict, C. R.: Carbon isotope discrimination in a plant possessing the C_4 dicarboxylic acid pathway. Biochem. Biophys. Res. Commun. *41*, 1205–1210 (1970)
Wilkins, M. B.: An endogenous rhythm in the rate of carbon dioxide output of *Bryophyllum*. I. Some preliminary experiments. J. Exp. Bot. *10*, 377–390 (1959)
Wilkins, M. B.: The effect of light upon plant rhythms. Cold Spring Harb. Symp. *25*, 115–129 (1960)
Wilkins, M. B.: An endogenous rhythm in the rate of carbon dioxide output of *Bryophyllum*. III. The effects of temperature changes on the phase and period of the rhythm. Proc. R. Soc. *156*, 220–241 (1962a)

Wilkins, M. B.: An endogenous rhythm in the rate of carbon dioxide output of *Bryophyllum*. IV. Effect of intensity of illumination on entraitment of the rhythm by cycles of light and darkness. Plant Physiol. *37*, 735–741 (1962 b)

Wilkins, M. B.: An endogenous rhythm in the rate of carbon dioxide output of *Bryophyllum*. V. The dependence of rhythmicity upon aerobic metabolism. Planta *72*, 66–77 (1967)

Wilkins, M. B.: Circadian rhythms in plants. In: Physiology of plant growth and development. Wilkins, M. B. (ed.), pp. 647–671. London: McGraw-Hill 1969

Wilkinson, M. J., Smith, H.: Properties of phosphoenol pyruvate carboxylase from *Bryophyllum fedtschenkoi* leaves and fluctuation in carboxylase activity during endogenous rhythms of carbon dioxide output. Plant Sci. Lett. *6*, 319–324 (1976)

Willert, D. J. von: The effect of NaCl on the accumulation of malate in leaf slices of the halophytic plant *Mesembryanthemum crystallinum*. Z. Pflanzenphysiol. *76*, 44–50 (1975 a)

Willert, D. J. von: Stomatal control, osmotic potential and the role of inorganic phosphate in the regulation of the crassulaceen acid metabolism in *Mesembryanthemum crystallinum*. Plant Sci. Lett. *4*, 225–229 (1975 b)

Willert, D. J. von: Die Bedeutung des anorganischen Phosphates für die Regulation der Phosphoenolpyruvatcarboxylase von *Mesembryanthemum crystallinum* L. Planta *122*, 273–280 (1975 c)

Willert, D. J. von, Brinckmann, E., Schulze, E. D.: Ecophysiological investigations in the coastal desert of southern Africa. Proc. 1st Europ. Ecol. Symp. Blackwell Scient. Publ., in press (1978)

Willert, D. J. von, Curdts, E., Willert, K. von: Veränderungen der PEP-Carboxylase während einer durch NaCl geförderten Ausbildung eines CAM bei *Mesembryanthemum crystallinum*. Biochem. Physiol. Pflanz. *171*, 101–107 (1977 a)

Willert, D. J. von, Kirst, G. O., Treichel, S., Willert, K. von: The effect of leaf age and salt stress on malate accumulation and phosphoenolpyruvate carboxylase in *Mesembryanthemum crystallinum*. Plant Sci. Lett. *7*, 341–346 (1976 a)

Willert, D. J. von, Kluge, M.: Studies on malate fluxes in leaf slices of *Bryophyllum daigremontianum*. Plant Sci. Lett. *1*, 391–397 (1973)

Willert, D. J. von, Kramer, D.: Feinstruktur und Crassulaceen-Säurestoffwechsel in Blättern von *Mesembryanthemum crystallinum* während natürlicher und NaCl induzierter Alterung. Planta *107*, 227–237 (1972)

Willert, D. J. von, Thomas, D. A., Lobin, W., Curdts, E.: Ecophysiological investigations in the family of *Mesembryanthemaceae*. Oecologia *29*, 67–76 (1977 b)

Willert, D. J. von, Treichel, S., Kirst, G. O., Curdts, E.: Environmentally controlled changes of phosphoenolpyruvate carboxylases in *Mesembryanthemum*. Phytochemistry *15*, 1435–1436 (1976 b)

Willmer, C. M., Johnston, W. R.: Carbon dioxide assimilation in some aerial plant organs and tissues. Planta *130*, 33–37 (1976)

Willmer, C. M., Pallas, J. E. Jr.: A survey of stomatal movements and associated potassium fluxes in the plant kingdom. Can. J. Bot. *51*, 37–42 (1973)

Winter, K.: CO_2 Fixierungsreaktionen bei der Salzpflanze *Mesembryanthemum crystallinum* unter variierten Außenbedingungen. Planta *114*, 74–85 (1973)

Winter, K.: NaCl-indizierter Crassulaceen-Säurestoffwechsel bei der Salzpflanze *Mesembryanthemum crystallinum*. Abhängigkeit des CO_2-Gaswechsels von der Tag/Nacht-Temperatur und von der Wasserversorgung der Pflanze. Oecologia *15*, 383–392 (1974)

Winter, K., Lüttge, U.: Balance between C_3 and CAM pathway of photosynthesis. In: Water and plant life. Ecological Studies, Vol. 19. Lange, O. L., Kappen, L., Schulze, E. D. (eds.), pp. 323–332. Berlin, Heidelberg, New York: Springer 1976

Winter, K., Lüttge, U., Ball, E.: $^{14}CO_2$ dark fixation in the halophytic species *Mesembryanthemum crystallinum*. Biochim. Biophys. Acta *343*, 465–468 (1974)

Winter, K., Lüttge, U., Winter, E.: Seasonal shift from C_3 photosynthesis to Crassulacean Acid Metabolism in *Mesembryanthemum crystallinum* growing in its natural environment. Oecologia, in press (1978)

Winter, K., Troughton, J. H.: Photosynthetic pathway in plants of coastal and inland habitats of Israel and Sinai. Flora *167*, 1–34 (1978 a)

Winter, K., Troughton, J. H.: Carbon assimilation pathways in *Mesembryanthemum nodiflorum* under natural conditions. Z. Pflanzenphysiol. *88*, 153–162 (1978b)

Winter, K., Troughton, J. H., Evenari, M., Läuchli, A., Lüttge, U.: Mineral ion composition and occurrence of CAM like diurnal malate fluctuations in plants of coastal and desert habitats of Israel and the Sinai. Oecologia *25*, 125–143 (1976)

Winter, K., Willert, D. J. von: NaCl-induzierter Crassulaceensäurestoffwechsel bei *Mesembryanthemum crystallinum*. Z. Pflanzenphysiol. *67*, 166–170 (1972)

Wolf, J.: Beiträge zur Kenntnis des Säurestoffwechsels sukkulenter Crassulaceen. II. Mitteilung der Untersuchungen zwischen Sedoheptose und Äpfelsäure. Planta *26*, 516–522 (1937)

Wolf, J.: Beiträge zur Kenntnis des Säurestoffwechsels sukkulenter Crassulaceen. III. Stoffliche Zusammenhänge zwischen gärfähigen Kohlehydraten und organischen Säuren. Planta *28*, 60–86 (1938)

Wolf, J.: Beiträge zur Kenntnis des Säurestoffwechsels sukkulenter Crassulaceen. VI. Mitt.: Neuere Vorstellungen vom Chemismus des Säurestoffwechsels. Planta *37*, 510–534 (1949)

Wolf, J.: Der diurnale Säurerhythmus. In: Encyclopedia of plant physiology. Ruhland, W. (ed.), Vol. 12, pp. 809–889. Berlin, Heidelberg, New York: Springer 1960

Wong, S. C., Hew, C. S.: Diffusive resistance, titratable acidity and CO_2 fixation in two tropical epiphytic ferns. Am. Fern J. *66*, 121–123 (1976)

Wood, H. G., Werkmann, C. H.: The utilization of carbon dioxide by propionic acid bacteria. Biochem. J. *32*, 1262–1271 (1938)

Wood, W. M. C.: Organic acid metabolism in *Sedum praealtum*. J. Exp. Bot. *3*, 336–355 (1952)

Wynn, T., Brown, H., Campbell, W. H., Black, C. C. Jr.: Dark release of $^{14}CO_2$ from higher plants. Plant Physiol. *52*, 288–291 (1973)

Xolocotzi, E. H.: Mexican experience. In: Arid lands in transition. Dregne, H. E. (ed.). Am. Assoc. Adv. Sci. Publ. *90*, 317–344 (1970)

Yamazaki, R. K., Tolbert, N. E.: Malate dehydrogenase in leaf peroxisomes. Biochem. Biophys. Acta *178*, 11–20 (1969)

Yoshida, A.: Enzymatic properties of malate dehydrogenase of *Bacillus subtilis*. J. Biol. Chem. *240*, 1118–1124 (1965)

Zabka, G. G., Chaturvedi, S. N.: Water conservation in *Kalanchoe blossfeldiana* in relation to carbon dioxide dark fixation. Plant Physiol. *55*, 532–535 (1975)

Zabka, G. G., Edelman, J.: Pre-illumination as a factor in the dark-fixation of CO_2 by *Kalanchoe blossfeldiana*. Plant Physiol. *36* (1961 a)

Zabka, G. G., Edelman, J.: Pre-illumination as a factor in the dark-fixation of carbon dioxide by leaf disks of *Kalanchoe blossfeldiana var*. Feuer Blüte. J. Exp. Bot. *13*, 11–19 (1961 b)

Zabka, G. G., Gregory, F. G., Edelman, J.: Dark fixation of carbon dioxide in *Kalanchoe blossfeldiana* in relation to photoperiodism. Nature (London) *813*, 1375 (1959)

Zabka, G. G., McMahon, E.: Relationships among CO_2 dark-fixation succulence, flowering, and organic acid formation in *Kalanchoe blossfeldiana* var. Tom Thumb. Can. J. Bot. *3*, 447–452 (1965)

Ziegler, H., Osmond, C. B., Stichler, W., Trimborn, P.: Hydrogen isotope discrimination in higher plants: Correlations with photosynthetic pathway and environment. Planta *128*, 85–92 (1976)

Zschoche, W. C., Ting, I. P.: Purification and properties of microbody malate dehydrogenase from *Spinacia oleracea* leaf tissue. Arch. Biochem. Biophys. *159*, 767–776 (1973)

Appendix

After the manuscript of this book was submitted to the publisher, some important recent papers became available to us. They are briefly reviewed here.

Queiroz provided a new review on the relevance of endogenous rhythms of enzyme capacity in CAM regulation [Queiroz, O.: CAM: rhythms of enzyme capacity and activity as adaptive mechanisms; to appear in Photosynthesis, Vol. II, Gibbs, M., Latzko, E. (eds.), Encyclopedia of Plant Physiol. (New Series), Berlin-Heidelberg-New York: Springer]. Regulation of glycolysis in CAM was studied by Pierre and Queiroz (Pierre, J. N., Queiroz, O.: Regulation of glycolysis and level of the crassulacean acid metabolism. Planta, in press). In this paper, the above authors argue that the allosteric regulation of key CAM enzymes cannot provide a complete explanation for the diurnal control of glycolysis in CAM. Studying *K. blossfeldiana*, where CAM can be induced by short days, Pierre and Queiroz found different patterns of glycolysis and different control steps depending on the actual CAM capacity. Under long-day treatment (i.e., low CAM capacity), all glycolytic enzymes and PEP-C were found to show a 12-h rhythm as result of superposition of two 24-h rhythms operating out of phase. Here phosphofructokinase is believed to be the main control point. Under short-day conditions, which increase the CAM capacity, there was an increase in the activity of the glycolytic enzymes, resulting in an overall 24-h rhythm with a maximum during the night. Regulation of nocturnal CAM is believed to be now at three enzymatic steps, namely phosphofructokinase, 3-P-glyceraldehydedehydrogenase/3-P-glycerate-kinase, and PEP-carboxylase. It remains an open question if the above findings are applicable also to CAM plants other than *K. blossfeldiana.*

The most detailed study until now on purification and characterization of PEP carboxylase from a CAM plant was provided by Jones et al. (Jones, R., Wilkins, M. B., Coggins, J. R., Fewson, Ch., Malcolm, A. D. B.: Phosphoenolpyruvate carboxylase from the crassulacean plant *Bryophyllum fedtschenkoi* Hamet et Perrier: purification, molecular and kinetic properties. Biochem. J. *175*, 391—406, 1978. The above authors were able to purify PEP-C to homogeneity in milligram amounts. Cross-linking experiments using a series of bis-imido esters of increasing length suggest that the native PEP-C occurs as a tetramer having a molecular weight of 385,000 and behaves as a dimer of dimers. The dimer seems to behave as an independent, stable, and active species of PEP-C. No evidence could be found for isoenzymes or for sub-unit heterogeneity. The apparent $Km_{(PEP)}$ is reported by the above authors to be 0.7 mM. HCO_3^- was saturating at the concentration caused by the dissolved atmospheric CO_2. 30 mM HCO_3^- was inhibitory. The pH response curve of PEP-C showed a distinct double optimum at pH 5.8 and 7.8. Malate inhibition was significant and of the competition type. It increases sharply with decreasing pH. Concerning the regulation of CAM, the above authors argue: "... With the reservations already mentioned, we have little evidence that variations in the amounts of enzyme or structural modifications play a significant part in regulation, and the overriding type of control must therefore involve reversible conformation changes... A sensitive control mechanism could be provided by the combined effects of malate and H^+ concentrations on PEP-C activity since it is quite likely that both vary in parallel in the cytoplasm."

Two recent papers deal with the PEP-C from *Mesembryanthemum crystallinum*, a plant where CAM can be induced by water and salt stress (see Chap. 3—6). Winter and Greenway [Winter, K., Greenway, H.: Phosphoenolpyruvate carboxylase from *Mesembryanthemum crystallinum*. Its isolation and inactivation in vitro. J. Exp. Bot. *39*, 539—546 (1978)] report that PEP-C in desalted extracts was much more stable when extracted from fully developed than from expanding leaves. The authors assume that another enzyme might be responsible in the inactivation of PEP-C. Greenway et al. [Greenway, H., Winter, K., Lüttge, U.: Phosphoenolpyruvate carboxylase during development of crassulacean acid metabolism and during a diurnal cycle in *Mesembryanthemum crystallinum*. J. Exp. Bot. *29*, 547–559 (1978)] found that characteristics of PEP-C changed during the first 12 d of CAM induction by 400 mM NaCl. The ratio PEP-C activity at pH 7/PEP-C activity at pH 8.0 increased with developing CAM, and $Km_{(PEP)}$ decreased. There was a clear shift of PEP-C characteristics during the diurnal CAM cycle. In extracts made during the first hours of the light period (phase of malate consumption) PEP-C activity at pH 7.0 was very low but high at 8.0. In contrast, extracts obtained in the phase of malate synthesis (i.e., during the first hours of the night), PEP-C activity was high both at pH 7.0 and 8.0. The authors interpret their results as indicative that PEP-C may exist in two states. One state is believed to favor rapid malate synthesis and transport to the vacuoles (dark period). The second state may dominate during the day in order to avoid interference of continued malate synthesis during conversion of malate to carbohydrates.

New information on malate transport into the vacuole based on electrochemical investigations became recently available (Lüttge, U., Ball, E.: Electrochemical investigations of active malic acid transport at the tonoplast into the vacuoles of the CAM plant *Kalanchoe daigremontiana*. J. Membrane Biol., submitted).

The membrane potentials of cells in leaf slices of *K. daigremontiana* were measured both in darkness and light to be −200 mV on average. The metabolic inhibitors ICCP ($5 \cdot 10^{-6}$ M) and CN^- ($5 \cdot 10^{-3}$ M) caused reversible depolarization. The membrane potential was found to be independent of the malic acid level of the cells; and to be hyperpolarized by fusicoccin ($30 \cdot 10^{-6}$ M), a fungal toxin. Fusicoccin inhibited malic acid accumulation in intact phyllodia of *K. tubiflora*. However, the re-mobilization of malate was not affected. Using the Nernst equation, Lüttge and Ball calculated the electrochemical gradients for the ions resulting from dissociation of malic acid (H^+, H-malate$^-$, malate^{2-}). The calculations suggest active transport of H^+ and H-malate$^-$ from the cytoplasm into the vacuole. Malate^{2-} is believed to migrate passively across the tonoplast. The above data allow the hypothesis that the active malic acid accumulation in the vacuoles during CAM is primarily an active H^+ transport at the tonoplast coupled with passive cotransport of malate^{2-}, with subsequent formation of H-malate$^-$ due to the actual vacuolar pH according to the equation $2H^+ + \text{malate}^{-2} \rightarrow H^+ + \text{H-malate}^-$.

An enzymatic technique has been developed for the isolation of leaf cells from *Sedum telephium* (Spalding, M.H., Edwards, G.E.: Photosynthesis in enzymatically isolated leaf cells from the CAM plant *Sedum telephium* L. Planta *141*, 59—63 (1978)]. In the light the cells performed high rates of $^{14}CO_2$ incorporation (30—70 μmol CO_2 mg^{-1} chl. h^{-1}). Half-maximum saturation of $^{14}CO_2$ fixation was at 2 mM bicarbonate (i.e., 20 μM CO_2). In the dark $^{14}CO_2$ fixation was about 3 μmol CO_2 mg^{-1} chl. · h^{-1}. O_2 evolution in the absence of exogenously supplied bicarbonate was observed, and there is no doubt that this O_2 evolution is due to photosynthesis operating at the expense of malate-derived CO_2 (cf. Chap. 3). However, addition of bicarbonate to the cell suspension enhanced O_2 production. Hence, it is conceivable that oxygen evolution occurring at ambient CO_2 concentrations is the result of photosynthesis consuming simultaneously both endogenous and exogenous CO_2. This interpretation is supported by the rather negative $\delta^{13}C$ values found in *S. telephium* (cf. Bender et al., 1973).

There are now further investigations on the effects of light intensity on CO_2 exchange and CO_2 balance in a CAM plant [Nose, A., Shiroma, M., Miyazato, K., Murayama, S.: Studies on matter production in pineapple plants. I. Effects of light intensity in light period on the CO_2 exchange and CO_2 balance of pineapple plants. Jap. J. Crop Sci. Vol. *46*, 580—587 (1977)]. These studies are consistant with the generalisation outlined in Chapter 5. In particular, it was shown that, as the light intensity in the light period increases, the net efflux of CO_2 decreases, while the subsequent CO_2 influx rate in both the light period and the following night is enhanced. Furthermore, the increase of light intensity changed the overall CO_2 balance into a direction toward a larger matter production.

The hypothesis that in CAM plants the actual internal CO_2 concentration determines the stomatal aperture (Kluge and Fischer, 1967; Kluge 1968b; Chap. 5.3.2.2 of this book) has been clearly supported by recent investigations of Cockburn et al. (Cockburn, W., Ting, I.P., Sternberg, L.O.: Relationship between stomatal behaviour and internal carbon dioxide concentration in CAM plants. Plant Physiol. in press, 1979). High internal CO_2 concentrations (often in excess of 2%) were shown to occur during the day when malate is decarboxylated. The stomatal resistance was always high when the internal CO_2 concentration was high. Experiments which manipulated the internal CO_2 concentration or where a high external CO_2 concentration was applied gave results which also support the view that the stomatal movements in CAM plants are controlled by the CO_2 concentration in the substomatal spaces.

Taxonomic Index

Subject Index

Progress in Botany

Morphology · Physiology
Genetics · Taxonomy
Geobotany

Fortschritte der Botanik
Morphologie · Physiology
Genetik · Systematik
Geobotanik

Editors: H. Ellenberg, K. Esser, H. Merxmüller, E. Schnepf, H. Ziegler

Volume 37

With contributions by numerous experts

1975. 20 figures, 1 table.
XVII, 402 pages (94 pages in German)
ISBN 3-540-07504-6

Volume 38

With contributions by numerous experts

1976. 22 figures. XVII, 377 pages (95 pages in German)
ISBN 3-540-07966-1

Fortschritte der Botanik is now to be continued under the title *Progress in Botany.* This is because the major part of the text is now written in English, and will make the volumes accessible to a wider reading public. The character of the series has not changed. The objective is still to furnish concise and prompt reports on all subject areas of botany. This involves the selection of works that describe worthwhile factual or methodological advances.

Springer-Verlag
Berlin
Heidelberg
New York